Management of Change in Chemical Plants

Management of Change in Chemical Plants

Learning from case histories

R. E. Sanders

Butterworth-Heinemann Ltd
Linacre House, Jordan Hill, Oxford OX2 8DP

 PART OF REED INTERNATIONAL BOOKS

OXFORD   LONDON   BOSTON
MUNICH   NEW DELHI   SINGAPORE   SYDNEY
TOKYO   TORONTO   WELLINGTON

First published 1993

© Butterworth-Heinemann Ltd 1993

British Library Cataloguing in Publication Data
Sanders, R. E.
   Management of Change in Chemical Plants:
   Problems and Case Histories
   I. Title
   660.068

ISBN 0 7506 1135 9

Library of Congress Cataloguing in Publication Data
Sanders, R. E.
   Management of change in chemical plants: problems and case
   histories/R. E. Sanders.
   p.   cm.
   Includes bibliographical references and index.
   ISBN 0 7506 1135 9
   1. Chemical plants – Safety measures.   I. Title.
   TP155.5.S213
   660'.2804–dc20   92–30167
                                                         CIP

Composition by Genesis Typesetting, Laser Quay, Rochester, Kent
Printed and bound in Great Britain

Contents

1 Introduction   1
Influence on mankind of the chemical industry   1
A glance at the history of chemical manufacturing   1
The modern industrial chemical industry modifies our way of
    living   2
The chemical industry's excellent safety record   3
Plant employee safety versus life style choices   5
Another way of considering fatal risks   5
References   7

2 Good intentions   8
Siphoning destroys a tender tank   8
    Afterthoughts   11
A mega-vessel is destroyed during commissioning   13
    Afterthoughts   16
A water drain is altered and a reactor explodes   16
    Afterthoughts   17
Good intentions on certain pollution reduction projects lead
    to troubles   19
An air system is improved and a vessel blows up   19
    Afterthoughts   23
Concerns for safety on a refrigerated tank   23
    Afterthoughts   25
Beware of impurities, stabilizers, or substitute chemicals   25
    Afterthoughts   25
A gas compressor is protected from dirt, but the plant catches
    fire   26
    Afterthoughts   26
The lighter side   26
A review of good intentions   26
References   27

3   Changes made to prepare for maintenance   28
A tank vent is routed to a water-filled drum to 'avoid'
    problems   28
    Afterthoughts   29
Preparing to paint large tanks   29
Preparing a brine sludge dissolving system for maintenance   30
    What happened in the brine system?   32
    Follow-up action   32
Violent eruption from a tank being prepared for repairs   32
    Afterthoughts   33
An explosion while preparing to replace a valve in an ice
    cream plant   34
    Afterthoughts   34
A chemical cleaning operation kills sparrows, but improves
    procedures   35
Other cleaning, washing, steaming and purging operations   37
A tragedy when preparing for valve maintenance   38
    Afterthoughts on piping systems   38
A review of changes made to prepare for maintenance   39
References   40

4   Modifications introduced in maintenance activity   41
Reboiler passes the hydrotest and later creates a fire   41
A tank explodes during welding repairs after passing a
    flammable gas test   43
    A phenol tank's roof lifts as repairs are made   44
    Another country, another decade and a similar tank
        explosion   44
Filter cartridges are replaced and an iron-in-chlorine fire
    develops   45
Repairs to a pipeline result in another iron-in-chlorine fire   45
Repair activity to a piping spool result in a massive leak from
    a sphere   46
The Phillips 66 incident   47
A breathing air system on a compressed air main is repaired   50
A hidden blind surprises the operators   51
    Afterthoughts on the use of blinds   53
Other incidents in which failure to remove blinds created
    troubles   54
Poor judgement by mechanics allowed a bad steam leak to
    result in a minor explosion   55
The Flixborough disaster and the lessons we should never
    forget   57
Do piping systems contribute to major accidents?   58
Specific piping system problems   59
    An 8 inch pipeline ruptures and an explosion occurs –
        24 December 1989   59

A 2 inch high-pressure hydrogen line fails and the fire
    topples a reactor – 10 April 1989   59
An 8 inch elbow ruptures from internal corrosion and a
    blast results in worldwide feedstock disruptions   60
Piping design failure incidents – lack of remote operated
    emergency valves   60
An 8 inch line ruptures in Mexico City and over 500 people
    die   61
References   61

**5 The one-minute modifier – small quick changes in a plant can
    create bad memories**   63
Explosion occurs after an analyser is 'repaired'   63
Just a little of the wrong lubricant   63
Instrument air back-up is disconnected   64
An operator modifies the instrumentation to handle an
    aggravating alarm   64
A furnace temperature alarm is altered   64
  Afterthoughts   65
The wrong gasket material creates imitation icicles in the
    summer   65
Another gasket error   67
As compressed asbestos gaskets are phased out other leaks
    will occur   67
Other piping gasket substitution problems   68
New stud bolts fail unexpectedly   69
Hurricane procedures are improperly applied to a tank
    conservation vent lid   69
  Afterthoughts on damages to the tank   70
Painters create troubles   70
Pipefitters can create troubles when reinstalling relief valves   72
Another pipefitter's error   74
A cooling water system is safeguarded and an explosion
    occurs some months later   75
Lack of respect for an open vent as a vacuum-relieving device
    results in a partial tank collapse   75
Lack of respect for an open vent as a pressure relief device
    costs two lives   76
  Afterthoughts   77
The misuse of hoses can quickly create problems   78
  Some of the many unpublished errors created with hoses   78
  The water hose at the Flixborough disaster   79
  Hoses used to warm equipment   79
  Three-Mile Island incident involves a hose   80
  The Bhopal tragedy is initiated by use of a hose   80
  Afterthoughts on 'one-minute modifications'   81
References   81

6   Failure to consult or understand specifications   83

Failure to provide operating instructions costs $100 000 in
     property damage   83
   What did the investigators find?   86
Other thoughts on furnaces   86
Low-pressure tank fabrication specifications were not
     followed   87
Explosion relief on low pressure tanks   87
Piping specifications were not utilized   87
   A small piping modification – for testing the line – created a
        failure point   87
   Correcting piping expansion problems   89
   Failure to follow piping specifications as piping supports are
        altered   90
   Piping system substitutions damage a six-stage centrifugal
        compressor   90
   Substitute piping material installed – accelerated corrosion/
        erosion results in a large fire   93
Pump repairs endanger the plant – but are corrected in time
        to prevent newspaper headlines   94
   Normal maintenance on a brine pump   94
   A cast iron pump bowl is installed in the wrong service   94
Plastic pumps installed to pump flammable liquids   95
Weak walls wanted – but alternative attachments contributed
        to the damage   95
An explosion could have been avoided if gasket specifications
        were utilized   96
Surprises within packaged units   97
References   97

7   'Imagine if' modifications and practical problem solving   99

'Imagine if' modifications – do not exaggerate the dangers as
        you perform safety studies   99
   New fire-fighting agent meets opposition – 'could kill
        people as well as fires'   99
   A process safety management quiz   100
   New fibre production methods questioned   102
Practical problem solving   102
   Clever approaches to problem solving   102
   The physics student and his mischievous methods   103
   What is the best approach?   104
References   104

8   Programmes to address ageing in chemical plants   105

Introduction   105
Integrity assurance of the containment system   106
Corrosion under insulation   107
Inspecting pressure vessels, storage tanks and piping   108
   Inspection of pressure vessels and storage tanks   109

One chemical plant's pressure vessel management
    programme   111
Inspection of above-ground piping   113
An assurance programme for safety relief valves and safety
    critical instruments   114
Safety relief valve considerations   114
    'In-house' testing of safety relief valves   115
    Selection of safety relief valve testing equipment   116
    SRV testing and repair procedures   118
    How often is often enough when testing SRVs?   120
    Keep the SRV records straight   121
    Communications to equipment owners and management   121
Process safety interlocks and alarms   122
    Protecting process safety interlocks at a DuPont plant   122
    Testing safety critical process instruments at a DuPont
        plant   123
    Another company – a different emphasis on safety critical
        instrument systems   123
    Another approach to proof-testing in Louisiana   124
What instruments are considered critical?   126
    Prioritizing critical loops   127
    Proof-test frequencies   129
    Administering the critical instrument proof-test programme   129
Additional information on mechanical integrity   131
References   131

9   Properly managing change within the chemical industry   133
    Preliminary thoughts on plant modification control   133
    A backward glance at earlier chapters   134
    Some of the 1970s chemical plants' approaches to plant
        changes   135
    How are chemical plants addressing plant modifications
        during the 1980s and beyond?   136
    The Center for Chemical Process Safety   138
    New recommendations and new regulations   138
    An overview of training in a management of change
        programme   141
    A workable approach for reviewing proposed plant
        modifications   141
    How should the potential hazards be identified and
        evaluated?   148
    Adherence to good engineering practice   149
    Predictive hazard evaluation procedures   150
    Variances, exceptions, and special cases of change   152
    A procedure to address taking alarms, instruments, or
        shutdown systems out of service   152
    Safety critical instrument setting changes   154
    Changes in safety critical instrument test and equipment
        inspection frequencies   154

Approvals, documentation and auditing   154
  Approvals, endorsements and documentation   154
  Auditing the management of change programme   155
  Some generic management of change audit questions   156
Closing thoughts on a management of change policy   156
References   157

10 Sources of helpful information when considering modifications   158
The best five books in chemical process safety – from a
    process engineer's viewpoint   158
General chemical process safety books   159
Practical information in ageing of pressure vessels, tanks,
    piping and safety critical instruments   161
Fire and explosion references   161
Other helpful resources   161

Index   165

Acknowledgements

A number of people deserve thanks for encouraging me and helping me with this challenge. As an engineer within a chemical manufacturing facility, opportunities to write articles did not seem realistic to me. In the early 1980s, after I had submitted a rather primitive proposed technical paper, Bill Bradford encouraged me to draft a manuscript. My first technical paper was on the subject of plant modifications and was presented to the AIChE in 1982.

In 1983, Trevor A. Kletz asked me to help him teach an American Institute of Chemical Engineers Continuing Education Course. I was shocked and elated to be considered. It was a great opportunity to learn from this living legend in loss prevention. It has been educational and enjoyable ever since; he has become my teacher, my coach, and my friend.

I assisted Trevor Kletz in teaching a two-day course entitled 'Chemical Plant Accidents – A Workshop on Causes and Preventions'. We periodically taught the course for six years and then he encouraged me to consider writing this book on plant modifications. Jayne Holder, formerly of Butterworths, was extremely supportive with all my concerns and questions.

Before I got started, I was searching for help, and William E. Cleary, Jack M. Jarnagin, Selina C. Cascio and Trevor A. Kletz volunteered to support the project. Then the hard part came. Again Trevor Kletz and Jayne Holder encouraged me to get started.

I am grateful to Bill Cleary for his technical and grammatical critique, and to Selina Cascio for her skill in manuscript preparation, including endless suggestions on style and punctuation. Jack Jarnagin's drafting assistance provided the clear illustrations throughout the text. I am grateful to Trevor for his continuous support.

Also, thanks go to my wife, Jill, for both her patience and her clerical help, to my daughter Laura for proofreading and to Warren H. Woolfolk for his help on Chapter 8. Thanks go to Bernard Hancock, of the Institution of Chemical Engineers (UK) for his generous permission to use a number of photos to enhance the text. I also thank the management of PPG Industries – Chemicals Group, my employer, for their support. Finally, I appreciate the many contributors of incidents and photographs who because of the situation wanted to remain anonymous.

Preface

There are many good articles as well as recently published good books on chemical process safety. These texts offer sound advice on identifying chemical process hazard analysis, training, audits, handling the management of change, etc. But there are only a few people, such as Trevor A. Kletz, who offer many case histories which reinforce the fact that there must be a better way to handle our business.

Trevor Kletz encouraged me to write a book on plant modifications several years ago. At that time, we were working together teaching an American Institute of Chemical Engineers Continuing Education Course entitled 'Chemical Plant Accidents – A Workshop on Causes and Preventions.' It is hoped that this book in some way mimics Trevor Kletz's style of presenting clear interesting anecdotes which illustrate process safety concepts. Hopefully, these case histories can be shared with chemical process operators, pipefitters, welders, operating foremen and maintenance supervisors.

Just as I was completing this book, the United States Department of Labor – Occupational Safety and Health Administration (OSHA) issued a new law. This law, entitled 'Process Safety Management of Highly Hazardous Chemicals: Explosives and Blasting Agents; Final Rule', was enacted on 24 February 1992. The final rule was to become effective on 26 May 1992. I have attempted to interpret the section of the law which dealt with management of change, based upon 18 years experience in loss prevention. OSHA representatives may or may not agree with specific procedures that are shown in Chapter 9; they might choose additional approval steps or additional documentation.

The reader should be aware that all my experiences were within a major chemical plant with about a billion dollars of investment, 1600 employees and covering 250 acres of chemical plant. There were toxic gases, flammable gases, flashing flammable liquids, and combustible liquids, but there were no significant problems with combustible dusts and no significant problems with static electricity.

The information in this book came from a number of sources including stories from my experiences in the now defunct Louisiana Loss Prevention Association, from students in the AIChE's 'Chemical Plant Accidents' course, co-workers, friends and other literature. It is believed that the stories are reliable. The approaches and recommendations made on each case seemed appropriate; however, the author, editor and publisher specifically disclaim that compliance with any advice contained herein will make any premises or operations safe or healthful, or in compliance with any law, rule or regulation.

Note: Where reference is made to the ton and the gallon, U.S. measurements have been retained: 'ton' refers to the 'short ton' of 2000 lb; 'gallon' refers to 0.8327 of the British gallon.

Chapter 1

Introduction

Influence on mankind of the chemical industry

The chemical and petroleum-refining industries have modified our lives for the better. While reviewing the shortcomings and accidents of the industries we must keep in mind the benefits that surround us.

Few individuals in the developed world stop to realize how the chemical industry has modified every minute of their day. The benefits of the chemical industries are apparent from the time our plastic alarm clock tells us it is time to get up from our polyester sheets, and our polyurethane foam mattress. As our feet touch the nylon carpet, we walk towards a phenolic light switch which allows electrical current to safely pass through polyvinyl chloride insulated wires to illuminate our way to the bathroom sink to wash our face in sanitized water using a chemically produced soap (Taylor, 1986).

A glance at the history of chemical manufacturing

Since people first roamed the earth, they have been devising ways of changing things and trying to make their lives a little easier. In the broad sense, prehistoric people practised chemistry, beginning with the use of fire to produce chemical changes like burning wood, cooking food, and the firing of pottery and bricks (Taylor, 1957a).

The oldest of the major industrial chemicals in use today is soda ash. It seems to date back to 3000 to 4000 B.C., because beads and other ornaments of glass, which presumably were made with soda ash, were found in Egyptian tombs. It seems that natural soda ash was used as an article of trade in ancient Lower Egypt (Columbia-Southern Chemical Corporation, 1951).

From what we know today, even the earliest civilized people were aware of the practical use of alcoholic fermentation. The Egyptians and Sumerians made a type of ale before 3000 B.C. and the practice may have originated much earlier (Taylor, 1957a).

The Romans and Greeks before the Christian era cleaned clothes and woollen textiles by treading the material or beating it with stones or a wooden mallet in the presence of fuller's earth together with alkali, lye, or more usually ammonia in the form of stale urine. Roman fullers put out pitchers at the street corners to collect urine. As repugnant as it seems to many, it should be noted that stale urine was used for cleaning clothes from Roman times up to the nineteenth century (Taylor, 1957b).

There have been periods of little significant progress, as in Europe during the centuries before the year 1100. During the 900s people only lived for about 30 years. Food was scarce, monotonous, and often stale or spoiled, and homes offered minimal protection from the elements. Fewer than 20% of the Europeans during the Middle Ages ever travelled more than 10 miles (16 km) from the place they were born. The age that followed these bleak years brought forth a burst of inventiveness as people began to understand how science could take over some of their burdens (Groner *et al.*, 1972; *The World Book Encyclopaedia*, 1980a).

In Europe, the harvesting and burning of various seaweeds and vegetation along the seashore to create a soda ash type of product is one of the earliest examples of recorded industrial chemical manufacturing. This type of chemical processing was fairly widespread before modern recorded history. In fact, the Arabic name for soda, *al kali* comes from the word 'kali', which is one of the types of plants harvested for this early industrial chemical-producing activity. The desired product of this burned vegetation was extracted with hot water to form a brown-coloured lye (primarily sodium carbonate, common name soda ash) which was used to manufacture soap and glass. Soda ash is by far the oldest of the major industrial chemicals used today (Columbia-Southern Chemical Corporation, 1951).

During the 1600s and 1700s scientists laid the foundations for the modern chemical industry. Germany, France and England initially manufactured inorganic chemicals to preserve meat and other foods, to make gunpowder, to dye fabrics and to produce soap. In 1635, the first American chemical plant started up in Boston to make saltpetre for gunpowder and for the tanning of hides. (*The World Book Encyclopaedia*, 1980b).

The chemical industry was being formed as the Industrial Revolution began, but as late as 1700, only 14 elements had been identified. The early chemical manufacturing process development can be accredited to Nicolas LeBlanc, a physician to the Duke of Orleans, who outlined a method of making soda ash starting with common table salt. Dr LeBlanc was promised an award in 1783 and the Duke of Orleans provided sufficient funds to build a plant not far from Paris (Hou, 1933), but in 1793 the Duke of Orleans was guillotined by the French revolutionists and LeBlanc never received the award. Other soda plants sprang up in France, England, Scotland, Austria and Germany. (*Academic American Encyclopedia*, 1983).

The modern industrial chemical industry modifies our way of living

During the 1800s chemists discovered about half of the 100 known elements. After 1850, organic chemicals such as coal-tar dyes, drugs, nitroglycerin explosives, and celluloid plastics were developed and manufactured. World Wars I and II created needs for new and improved chemical processes for munitions, fibre, lightweight metals, synthetic rubber and fuels (*The World Book Encyclopedia*, 1980b). The 1930s witnessed the production of neoprene (1930), polyethylene (1933), nylon (1937) and glass fibre (1938) which signalled the beginning of an era that would see plastics replace natural materials. These 'plastics' would radically influence how things were designed, constructed and packaged (Industrial Risk Insurers, 1980).

By the 1950s and 1960s chemical processing had become more and more sophisticated, with larger inventories of corrosive, toxic and flammable chemicals. It became no longer acceptable for a single well-meaning individual to quickly change the design or operation of a chemical or petrochemical plant without reviewing the side effects of these modifications. Many case histories of significant accidents vividly show examples of narrowly focused individuals who cleverly solved a specific problem and failed to examine other consequences (Kletz, 1976, 1988a,b; Russell, 1976; Booth, 1976; Sanders, 1983; Sanders *et al.*, 1990).

This book will focus on a large number of near misses, damaging fires, explosions, leaks and injuries. A 'plant modification' was determined to be at least a contributory cause in each of these accidents. A reader who is not familiar with the chemical industry might be tempted to think that the chemical industry is one of the most hazardous industries. However, the opposite is true. The United States chemical industries (and most European chemical industries) are the safest of all industries.

The chemical industry generally handles business so well that it is difficult to find significant numbers of recent incidents. Many of the featured case histories occurred over 15 years ago; however, the lessons that can be learned will be appropriate into the 21st century. Tanks can fail from the effects of overpressure and underpressure in the 1990s just as well as they failed in the 1970s. Incompatible chemicals are incompatible in any decade and the human can be forgetful at any time. Before the first case histories are presented, I will discuss the safety record of the chemical industry.

The chemical industry's excellent safety record

The National Safety Council of the United States produces volumes of statistics on recordable work injuries and illnesses based on record keeping requirements of the US Occupational Safety and Health Act of 1970. The 1988 figures were based upon 3.4 million employees. (National Safety Council, 1990).

The National Safety Council studies show that the chemical manufacturing industries are in the top four safest performer ratings of 43 'principal industries' in terms of 'loss-of-time injuries'. The 1988 chemical industry had an incident rate of 0.59. This means that if a group of 1000 chemical workers worked 40 hours a week for a year (or 2000 hours each) then the chemical industry could expect about six individuals to experience a loss-of-time injury. By contrast, this is one-third of the 2.16 rate of cases involving days away from work and deaths in the 1988 wholesale and retail trade. The trucking industry experiences rates of 16.55, or about 27 times higher than the chemical industries. In short, the 16.55 figure suggests that if there were 1000 people in a crew in the trucking industry, 166 personnel would experience an accident that would keep them away from work (for at least one day) during the year (Table 1.1).

Table 1.1 1988 Incidence rates of principal industries

Textile	0.52
Chemical	0.59
Furniture and fixtures	1.63
Whole and retail trade	2.16
All industries average	2.17
Paper	2.42
Construction	3.32
Railroad transportation	5.04
Trucking	16.55

Incident rate per 100 full-time employees involving cases in which the employee lost time from work, including death. Data courtesy of the National Safety Council, Chicago, Illinois, USA.

The National Safety Council reports many other types of accident statistics. Another relevant index of the safety is the total recordable cases of injuries. Over 300 000 such recordable cases were compiled in 1988 and the chemical industry had a record that could be envied by many other industries.

For both 1987 and 1988 the chemical industry was reported to have a much safer recordable incident rate than 41 of the 43 industries in the Council's surveys. Only the communications industries and the oil and gas extraction industries had better experiences during those years.

The National Safety Council's fatal accident frequency rate is another measure of safety for which the chemical industry has historically had a good record. Fortunately, there are so few fatal accidents during any year that records from any one classification of any one year may or may not be typical. Using the nine work classifications as in the 'lost time incident frequency rates' and averaging them for the five years from the 1984 to 1988, we find that only the textile industry is safer (Table 1.2) (National Safety Council, 1986, 1987, 1988, 1989, 1990).

Table 1.2 Fatal accident incidence rates of principal industries

Textile	0.0023
Chemical	0.0032
Furniture and fixtures	0.0039
Wholesale and retail trade	0.0075
All industries average	0.0053
Paper	0.0057
Construction	0.0148
Railroad transportation	0.0123
Trucking	0.0237

Incident rate per 100 full-time employees. These are averages of 1983 to 1988 statistics. Data courtesy of the National Safety Council Chicago, Illinois, USA.

This indicates that if a chemical plant had 10 000 employees working for 10 years, one could expect about three employees to perish. By contrast, wholesale and retail trade occupations would experience about 2.3 times more fatalities and the trucking industry would experience about 7.4 times as many fatal accidents for the same number of people and same length of employment. However, the average person probably perceives that working for a chemical industry is a lot more dangerous that working for the wholesale and retail trade or for the trucking industry. Fires, explosions and releases of disagreeable gases make front pages of newspapers, and few people are aware of the high degree of safety within a chemical plant. The chemical industry must find better ways to educate the public on just how safe the industry really is.

Plant Employee Safety vs. Life Style Choices

The Chemical Manufacturers' Association, CMA, has published a 57-page booklet entitled *Risk Communication, Risk Statistics, and Risk Comparisons: A Manual For Plant Managers* (Covello *et al.*, 1988). It is a practical guide on chemical risks. It offers a number of concrete examples of risk comparisons and offers two pages of warnings on the use of such data. Among other things, the 'warning notes' suggest that the accuracy cannot be guaranteed; some of the data could be outdated, and typical risk data are a hodgepodge of risks characterized by different levels of uncertainty. However, this booklet offers about 13 tables or charts of very interesting comparisons.

The data in Table 1.3 form part of the CMA's booklet and appeared in Cohen and Lee (1979). Table 1.3 lists only 14 of the 448 causes.

Table 1.3 Estimated loss of life expectancy by lifestyles

Cause	Days sacrificed
Cigarette smoking (male)	2250
Being 30% overweight	1300
Being a coal miner	1100
Being 20% overweight	900
Cigarette smoking (female)	800
Cigar smoking	330
Dangerous jobs (accidents)	300
Motor vehicle accidents	207
Average jobs (accidents)	74
Drowning	41
Safest jobs (accidents)	30
Coffee	6
Oral contraceptives	5
Diet drinks	2

Another way of considering fatal risks

Some British authors from the chemical industry began discussing risks to employees using a concept of 'fatal accident frequency rates' (FAFR) in the early 1970s. The FAFR was developed to avoid the use of small fractions as in Table 1.2. This rate can be thought of as the number of deaths from industrial injury expected by a group of 1000 employees during a 50 year working career with the employee working 40 hours a week for 50 weeks a year (or the number of deaths expected per 100 000 000 employee working hours). Widely published articles of the time presented data as shown in Table 1.4 (Kletz, 1977). These numbers offer a good idea of the relative risks of working within the chemical industry versus working in the construction industry or driving a car.

Table 1.4 British fatal accident frequency rates

Industrial activities	
Clothing and footwear	0.15
Chemical industry	4
Average British industry	4
Steel industry	8
Deep sea fishing	35
Construction workers	67
Crews aboard aircraft	250[a]
Professional boxers	7000[a]
Non-industrial activities	
Staying at home (men, ages 16 to 65)	1
Staying at home (general population)	3
Travelling by train	5
Travelling by car	57
Motorcycling	660
Canoeing	1000

[a] While working
Data of 1960s and 1970s

More up-to-date information from the UK's Health and Safety Executive is available from data collected between 1974 and 1978 (British Medical Association, 1987). It can be converted to fatal accident frequency rates and compared to the US National Safety Council's figures collected between 1983 and 1988 (Table 1.5).

Table 1.5 Fatal accident frequency rates comparing the United Kingdom and the United States (National Safety Council, 1985, 1986, 1987, 1988, 1989; British Medical Association, 1987)

	UK	US
Manufacture of clothing and footwear	0.25	NA
Manufacture of textiles	NA	1.2
Manufacture – timber, furniture, etc.	2.0	NA
Manufacture of furniture and fixtures	NA	2.0
Chemical manufacturing	4.3	1.6
Construction	7.5	7.5
Railway transportation	9.0	6.0
Trucking – local and long distance	NA	11.8

The British data are a minimum of five years older than those from the United States and on the average they are about ten years older. This compilation compares the relative safety of the chemical industry to that of other industries, such as construction and transportation, using up-to-date FAFR. This relative safety must be kept in mind as you review the case histories in the rest of this book.

References

Academic American Encyclopedia (1983) Vol. 4. 'Chemical Industry and History of Chemistry', Grolier Inc., Danbury, CT, pp. 317–325

Booth, G. (1976) Process changes can cause accidents. *Loss Prevention,* American Institute of Chemical Engineers, **10**, pp. 76–78

British Medical Association (1987) *Living with Risk*, Wiley, Chichester, p. 68

Cohen, B. and Lee, I. (1979) A catalog of risks. *Health Physics,* **36**, pp. 707–722

Columbia-Southern Chemical Corporation (1951) *Soda Ash*, Columbia-Southern Chemical Corporation – subsidiary of Pittsburgh Plate Glass Co., Pittsburgh, pp. 3–7

Covello, V.T., Sandaman, P.M. and Slovic, P. (1988) *Risk Communication, Risk Statistics, and Risk Comparisons: A Manual for Plant Managers*, Chemical Manufacturers' Association, Washington D.C.

Groner, A. and the Editors of American Heritage and Business Week (1972) *The American Heritage History of American Business and Industry*, American Heritage Publishing Co. Inc., New York, pp. 10–21

Industrial Risk Insurers (1980) *The Sentinel*, IRI, Hartford, CT, October–November, pp. 4–5

Kletz, T.A. (1976) A three-pronged approach to plant modifications. *Chemical Engineering Progress* (November), pp. 48–55

Kletz, T.A. (1977) Evaluate risk in plant design. *Hydrocarbon Processing*, **56**, pp. 297–324

Kletz, T.A. (1988a) Modifications. In *What Went Wrong? Case Histories of Process Plant Disasters*, 2nd edn, Gulf Publishing, Houston, Chapter 2

Kletz, T.A. (1988b) *Learning from Accidents in History*, Butterworth Scientific, Guildford, UK, Chapters 1, 7, 8, 14

National Safety Council (1986) *Work, Injury and Illness Rates 1985*, NSC, Chicago, IL

National Safety Council (1987) *Work, Injury and Illness Rates 1986*, NSC, Chicago, IL

National Safety Council (1988) *Work, Injury and Illness Rates 1987*, NSC, Chicago, IL

National Safety Council (1989) *Work, Injury and Illness Rates 1988*, NSC, Chicago, IL

National Safety Council (1990) *Work, Injury and Illness Rates 1989*, NSC, Chicago, IL

Russell, W.W. (1976) Hazard control of plant process changes. *Loss Prevention,* **10**, pp. 80–87

Sanders, R.E. (1983) Plant modifications: troubles and treatment. *Chemical Engineering Progress* (February), pp. 73–77

Sanders, R.E., Haines, D.L. and Wood, J.H. (1990) Stop tank abuse. *Plant/Operations Progress* (January), pp. 61–65

Taylor, F.S. (1957a) *A History of Industrial Chemistry*, W. Heinemann Ltd, London, pp. 21, 59

Taylor, F.S. (1957b) *A History of Industrial Chemistry*, W. Heinemann Ltd, London, pp. 153–155

Taylor, F.S. (1957c) *A History of Industrial Chemistry*, W. Heinemann Ltd, London, p. 130

Hou, Te-Pang, (1933) *Manufacture of Soda with Special Reference to the Ammonia Process – A Practical Treatise*, The Chemical Catalogue Company, American Chemical Society, New York, pp. 15–17

The World Book Encyclopedia (1980a) Vol. 13. 'Middle Ages'. World Book – Childcraft International Inc., Chicago, IL, p. 432

The World Book Encyclopedia (1980b) Vol. 3. 'Chemical Industry'. World Book – Childcraft International Inc., Chicago, IL, pp. 310–314

Chapter 2

Good intentions

Chemical plant modifications are essential for survival in the dynamic chemical manufacturing industry. The goal of these modifications may be to improve yields, to compensate for unavailable equipment, to increase production, to add storage capacity, to reduce costs, to improve safety, and/or to reduce pollution potentials. The means to achieve these goals may be changes in piping or equipment, new operating procedures, new operating conditions, changes in material of construction, as well as the process chemical changes in feedstocks, catalysts, fuels or their method of delivery.

The first series of modifications featured were all 'good intentions'. In spite of creative ideas and considerable efforts these modifications failed because no one took the time to examine and expose the weaknesses. These undetected weaknesses caused undesired side effects.

Siphoning destroys a tender tank

A chemical plant complex was designed to use high volumes of brackish water from an adjacent river for once-through cooling. Occasionally, trace emissions of caustic soda were present in the effluent river water and the company was concerned with pH excursions. An engineer was assigned to make modifications to improve the situation.

The engineer designed a system that would allow enough hydrochloric acid into the effluent to neutralize the caustic traces to form more environmentally acceptable table salt. A lightweight 'off the shelf' glass fibre tank about 8 ft (2.2 m) in diameter and 8 ft (2.2 m) high was installed with piping and controls. The purpose of this modification was to receive, store and meter out acid to control the pH of the effluent. The tank was an atmospheric closed-top design and had two top nozzles.

During normal operation the vessel would receive acid via an in-plant pipeline. It was also equipped to receive acid via tank trucks (Figure 2.1). One of the top nozzles was the fill line and the other was piped to a small vent scrubber to eliminate fumes given off during the filling operation.

The intentions to improve the environment were noble, but the simple vent system design possessed an unrecognized flaw which allowed a minor overfill situation to ruin the vessel. This was not a complex project, so the

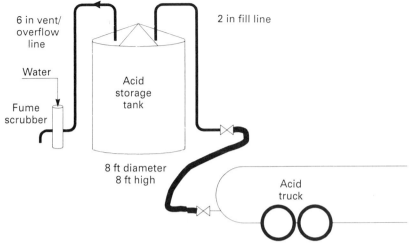

Figure 2.1 Acid tank just before failure (Courtesy of J. M. Jarnagin)

proposal was accepted and no one asked pointed questions like 'What happens if the acid storage tank is overfilled by a tank truck?' (As we shall see, many unforeseen problems have occurred on equipment so simple that a detailed examination of the proposal was considered unnecessary.)

About a year after the system was put in service, the tank was filled via a tank truck, instead of the usual pipeline. As the tank truck was unloaded the acid level rose in the small storage tank. The tank was overfilled and acid started to overflow through the 6 inch (15 cm) line into the scrubber. The alert truck driver responded by abruptly shutting the delivery valve from his truck. Unexpectedly, the partial vacuum created by the siphoning action of the overflowing liquid, exceeded the tank's vacuum rating and the storage tank was totally destroyed (Figures 2.2, 2.3 and 2.4).

Figure 2.2 Tank truck parked next to acid tank

Figure 2.3 Acid tank just after failure

Figure 2.4 Close-up of acid tank damage

Afterthoughts

Storage tanks are more fragile than many experienced operations, engineering and maintenance employees realize. The Institution of Chemical Engineers, Warwickshire, UK, offers a training kit with slides, booklets and instructor's guides for in-plant training. The kit is called 'Hazards of Over – and Under – Pressuring of Vessels' and it is highly recommended (Institution of Chemical Engineers, Hazard Workshop Module 001).

Trevor Kletz created this hazard workshop, and it is written in easy-to-understand terms. It is excellent for chemical process operators, chemical plant supervisors, and process engineers. It explains the strength of low-pressure storage tanks in clever ways, including the information in Table 2.1, which was based upon one of the slides in the kit. Table 2.1 was copied with permission from Trevor Kletz.

Table 2.1 Many atmospheric and low-pressure tanks appear big and strong. But don't believe it! Think about a can of baked beans and compare the pressure-resisting strength of the can's shell and lid to that of a low-pressure storage tank. If a baked bean can was assigned a strength of one, then:

	Shell	Roof
A small can of baked beans	1	1
A standard drum		
40 gallon UK	1/2	1/3
55 gallon US		
100 cubic metre tank (15 ft diameter, 20 ft high)	1/4	1/11
1000 cubic metre tank (40 ft diameter, 30 ft high)	1/8	1/57

Obviously, a multidisciplinary safety review committee may have detected the problems of the ill-advised use of a combination vent/overflow line, but this type of collapse is still considered somewhat of an oddity. Several variations of this type of collapse have been recently reported on low-pressure and atmospheric tanks. One company accidentally experienced two such partial collapses in 1989. Another company reported an incident in which operators were asked to fill a tank as high as possible for storage needs, so they ignored the high-level alarm. After the tank level reached the overflow, the liquid started pouring out of the overflow faster than it was being pumped in and the tank collapsed (T.A. Kletz, personal communication).

Every low-pressure, closed-top tank with an overflow line should be equipped with some type of device to prevent collapse from an overfilling operation. A large open nozzle, a high-level alarm and/or shutdown system, vacuum breakers or separate vent lines are some appropriate safeguards.

Another point to consider is that the acid tank had been filled above the upper sight glass nozzle and the operator mistakenly allowed the transfer to continue. Oddly enough, but I do not recall ever seeing the maximum filling level on a low pressure tank drawing or a simple guideline on the acceptable range for filling tanks (as Figure 2.5), perhaps because most low-pressure tanks are equipped with both vents and overflow lines. Maybe most people feel such a sketch is just too elementary, but the truth is that many tanks are damaged by overfilling.

In a different incident, a low-pressure tank 100 ft (30 m) in diameter and over 40 ft (12 m) high was distorted and the top ring split as a result of overfilling (Figures 2.6 and 2.7). The tank dome was designed for a maximum of 3 inches (0.75 kPa gauge) water pressure. However, an inattentive operator allowed liquid to rise over a foot into the dome (or about four times the rating) prior to the split of the roof to wall seam and the resulting spill.

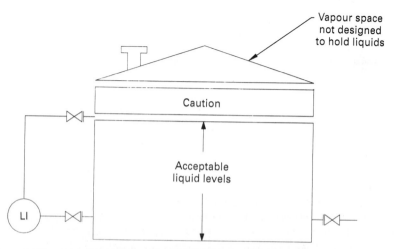

Figure 2.5 Low-pressure storage tank levels (Courtesy of J. M. Jarnagin)

Figure 2.6 Liquid product spilling from the damaged tank

Figure 2.7 Large tank (100 ft diameter) damaged by overfilling

A mega-vessel is destroyed during commissioning

A major petroleum refinery was expanding. Four new pressure vessels, called coker drums, were designed, fabricated and installed. These vessels were massive and very visible from the highway.

The coker drums were 27 ft (8.2 m) in diameter, 74 ft (23 m) tangent to tangent and 105 ft (32 m) in overall height. They were designed for 55 psig (380 kPa gauge) pressure with the weight of a complete charge of coke. They were not designed for a full vacuum. The wall thickness of the lower shell was 0.836 inches (2.1 cm).

Prior to start-up, these vessels were steamed out for a steam test to check for system leaks and to displace any oxygen. A modification, temporary piping 8 inches (20 cm) in diameter, was installed to the pre-engineered 24 inch (61 cm) vent to release the steam to the atmosphere (Figure 2.8). Unfortunately, the design of the piping allowed a water trap within a loop to collect water as the steam condensed.

The B Unit was steamed out first and vented through the 8 inch steam vent piping. The failure occurred about four days later. Witnesses said the coker appeared to crush-in like an aluminium beer can being squeezed in the middle (Figures 2.9 and 2.10). The vessel tore away from levels of decking and was impractical to salvage.

Further study revealed that the A Unit was being steamed out just after the steam was blocked to the B Unit. The steam in the B Unit continued to condense as the unit cooled while steam continued in the A Unit for two additional days. This additional condensate supply is believed to have been the source of the water which filled a 27 ft (8.2 m) section of the 24 inch (61 cm) line. The water column estimate is based upon the fact that the internal temperature of the B Unit was 144°F.

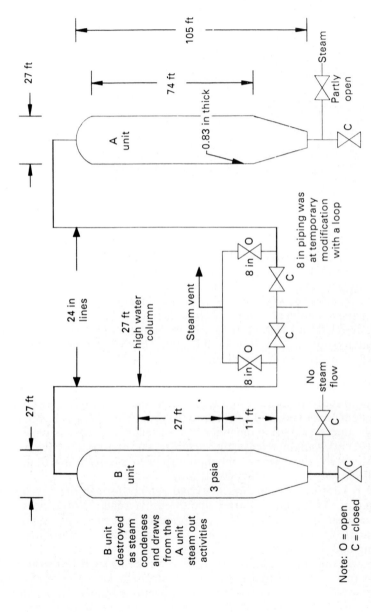

Figure 2.8 Conditions at time of failure of coker drum

Figure 2.9 Damaged coker drum in the unit (Courtesy of S. O. Kapenieks)

Figure 2.10 Close-up of the destroyed coker drum (Courtesy of S. O. Kapenieks)

Afterthoughts

Water has the unique ability to expand 1570 times when changing from a liquid to a vapour, and to contract 1570 times when it is condensing. This characteristic has made it an excellent working fluid for steam engines and turbines. But these properties can result in vacuum damage as just shown and explosive overpressure problems as presented in the next example of a 'good intention'.

A water drain line is altered and a reactor explodes

The supervisor of an organic chemical unit was concerned about fugitive emissions of an organic from a small-diameter water drain line. He no doubt felt that with just about 30 ft (9 m) of small 1½ inch (3.8 cm) piping, a couple of valves and a small conical bottom tank, this heat transfer fluid could be recovered. The aim was to reduce pollution potential and waste water treatment costs. The supervisor probably only focused on an area of concern, as in Figure 2.11. However, this small change in 'just a cooling water system' contributed to the destruction of a 110 psig (760 kPa gauge)-rated 9 ft (2.28 m) diameter pressure vessel. A short-lived fire ensued. The severe overpressure damage to this highly engineered

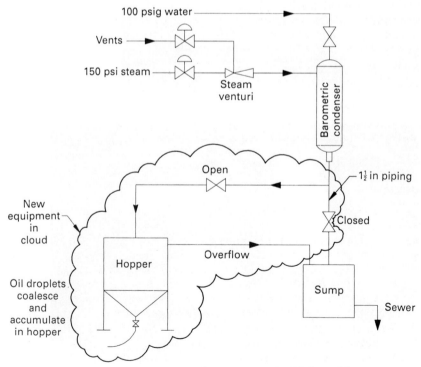

Figure 2.11 Modifications to reduce pollution (Courtesy of J. M. Jarnagin)

nickel-clad reactor resulted in over $500 000 (US 1987) in property damages (Sanders *et al.*, 1990).

The modification involved the rerouting of the discharge water from the barometric condenser. Rather than being discharged into a chemical sewer via a sump, it had been routed to a hopper which allowed the entrained heat transfer fluid droplets to drop out of solution and be recovered. Decontaminated water overflowed to the chemical sewer (Figure 2.11).

Each of the five parallel reactors was similarly modified, and they operated for over two years without incident (Figure 2.12). Heat transfer fluid was regularly recovered and the modification was deemed a success. But, on a rather cool December day, it became necessary to perform maintenance on the hopper. Cooling water from the barometric condensers of all the reactors was routed to the sump rather than to the hopper.

Shortly after the valve to the sump was opened and the valve to the hopper was closed, the no. 4 reactor began experiencing erratic pressure problems. The relief valve on the reactor cooling chest opened after about 15 minutes of system pressure swings and increases as the foreman and operator were troubleshooting the problem. About 20 seconds later a large section of a 9 ft (2.7 m) in diameter expansion joint burst, and the resulting thrust shifted the massive reactor.

A cloud of hot combustible heat transfer fluid jetted out of the damaged vessel, and the fire water deluge system was activated. Approximately 10 minutes later a 15–20 minute fire occurred. The operating and the emergency squad responded well and quickly extinguished the burning heat transfer oil that had sprayed over the adjacent piping and vessels. Two individuals experienced eye irritations but fortunately there were no serious injuries.

The investigating team concluded that the drain line to the sump was plugged or partially plugged, perhaps as a result of frozen heat transfer fluid above the block valve to the sump. Either well water or steam condensate backed into the hot oil system. The explosive vaporization of the water created pressures in excess of 600 psig (4100 kPa gauge) in a system designed for just 110 psig (760 kPa gauge).

Afterthoughts

The operations team in this area of the plant was convinced that if changes were to be made in any equipment, reactants, intermediates, catalyst, or any operating conditions of the process, the system would be first reviewed by a committee. But in the operations team's mind, this was just a waste water line. As apparent in Figure 2.12, just blocking the water drain lines at any time the system is in operation would cause physical explosion. Before the plant was restarted, all five barometric condensers were inspected to determine if there were any calcium deposits in the drain lines. The barometric condenser drain piping was simplified by routing it directly to the hopper without a valve and without a branch connection. A long-term study was initiated to determine if vacuum engines should be used as an alternative to barometric systems.

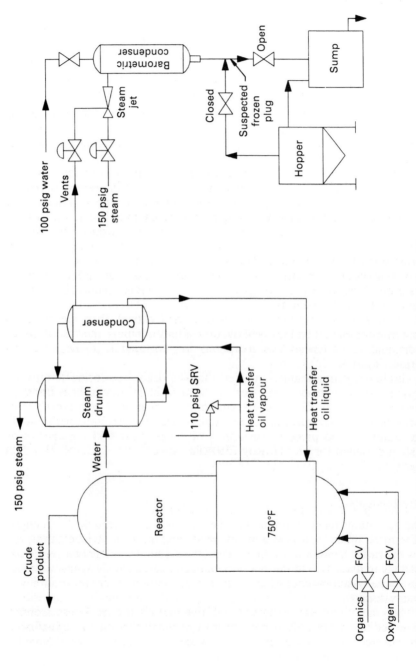

Figure 2.12 Solvents reactor system just before explosion (Courtesy of J. M. Jarnagin)

Good intentions on certain pollution reduction projects lead to troubles

Bodurtha (1976) says

'It is becoming increasingly evident that explosions and fires are occurring with greater frequency with pollution control equipment . . . The responsibility for safe design and operation rests with industry. Here, proper recognition of the hazard of the *total* system and not simply the pollution control device by itself is essential to safety.'

This advice was given out in 1976 but was not heeded in the recent accident just described.

Bodurtha also points out that the attempts to reduce the formation of oxides of nitrogen by the reduction of excess air in furnaces can result in the system developing fuel-rich mixtures. Such fuel-rich mixtures develop into flammable concentrations when air is added and explosions can occur.

Kletz reported that at the insistence of environmental authorities, the fumes of a benzene tank were required to be chemically treated within a scrubber. The benzene tank was heated to keep the benzene from freezing at 5°C (41°F). The scrubber was taken out of service during a plant shutdown. However, the steam heating system on the benzene tank was not temperature controlled so it continued to heat up and some of the benzene vaporized and condensed in the scrubbing system (T.A. Kletz, personal communication).

When the scrubber circulation system was resumed, there was a violent reaction between the liquid benzene and the chemical used to scrub the vapours. The scrubber was completely destroyed and the largest pieces were less than 1 ft² in area.

William Doyle reported some years ago, that one group within the US Environmental Protection Agency on the California coast insisted on modifications to reduce gasoline pollution at gasoline service stations. The modifications involved the fan-assisted recovery of gasoline vapours while gasoline was being pumped into vehicle fuel tanks. According to his sources there were over 20 accidental fires in those systems in the first four months of operation (Doyle, 1976).

An air system is improved and a vessel blows up

Talented supervisors often need to make improvements, but with their zeal to improve they may fail to see the need for a committee review for a small change in a utility system. In a previous example, a small change was made in 'just a waste water line' and damages of over a half-million dollars were experienced.

This case involves a 50 psig (345 kPa gauge) air system. It was probably easy to justify not initiating a committee review for decreasing moisture within a 40-year-old compressed air system, but it is easy to see the narrow focus of the innovators now.

Two air dryers contained activated alumina desiccant and moist compressed air was routed through one dryer until it was ready to be

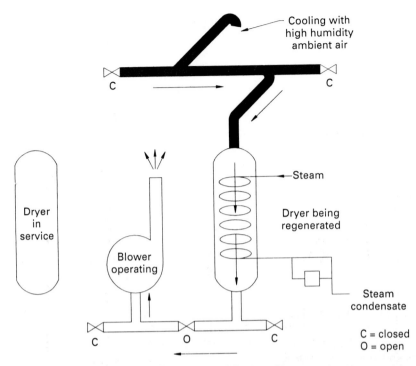

Figure 2.13 Air dryer regeneration before modifications (Courtesy of J. M. Jarnagin)

regenerated. The moist air was then switched to the companion dryer, and the spent dryer was heated by internal steam coils until the dryer achieved a certain temperature. The moisture was drawn out of the 'hot' dryer by the blower. However, the blower was sucking in ambient air from a very humid environment (Sanders *et al.*, 1990) (Figure 2.13).

This compressed air system had supplied air with an undesirable moisture content for a number of years. For several years prior to this incident this small air system was idle.

Innovators decided to use compressed, dry air to purge out the moisture and cool down the 'hot' unit after it achieved the appropriate temperature. They first used hoses to connect to the dry air system and this 'improvement' led to better dewpoints and longer dryer cycles. The temporary hose system seemed so beneficial that it was replaced with piping (Figure 2.14).

One evening, just about 5 minutes after a dryer was switched from the heating to the cooling cycle, the dryer exploded. The 30 inch (76 cm) diameter air dryer top head suddenly tore from the shell and the explosive release of air violently propelled activated alumina desiccant throughout the building. A large window panel about 6 ft (1.8 m) by 8 ft (2.4 m) blew out of the brick wall of this large building into a road. Luckily no one was

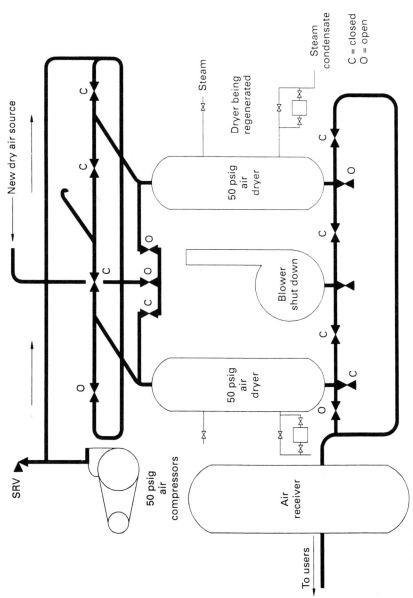

Figure 2.14 Modified regeneration sequence at time of explosion (Courtesy of J. M. Jarnagin)

Figure 2.15 Employee examines the air dryer failure

in the building when the vessel ruptured! There were no injuries. (See Figures 2.15 and 2.16.)

These creative individuals had set up a scheme to purge through the top of the air dryer and out of a partially opened ball valve on the piping beneath. They no doubt did not realize that the dryers were not individually equipped with safety relief valves. The overpressure protection was on the compressors. A new operator to the area did not open the exit valve sufficiently during the cool down step.

This chemical processing unit had three different air compressor systems operating at three different pressures, 50 psig (345 kPa gauge), 100 psig (690 kPa gauge), and 250 psig (1724 kPa gauge) for specific needs. In short, it did not occur to the innovators that they were tying a 250 psig (1724 kPa gauge) compressed air system into an air system with equipment that was designed for 50 psig. The apparent success of the modification blinded them to the additional risks that they created.

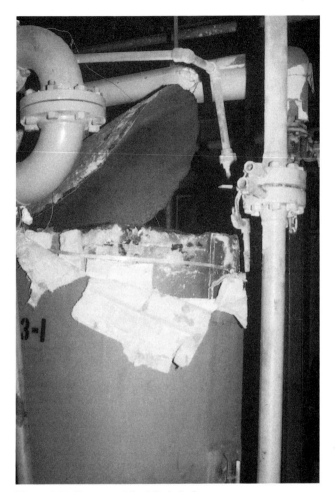

Figure 2.16 Close-up of the failed air dryer

Afterthoughts

This equipment was over 40 years old and was purchased by the previous owner. Documentation was very limited. The dryers were not even protected from the possibility of a high-pressure steam coil leak. After this incident the entire system was dismantled and scrapped.

Concerns for safety on a refrigerated tank

A large atmospheric liquid ethylene tank was constructed within a chemical complex. This tank was equipped with a single refrigeration unit to maintain the −155°F (−101°C) that was required to keep the vapour pressure below the 1.5 psig (10.4 kPa gauge) setting of the safety relief valve. In the event of problems with the refrigeration unit, the design concept was to allow the safety relief valve to open and the auto-refrigeration of the escaping vapour to cool down the remaining liquid in the tank (Kletz, 1974).

While construction was in progress, a process safety review determined that during refrigeration unit outages, there could be problems. On days when the safety relief valve opened, and there was little wind, the cold gases could drift down to ground level where they might find a source of ignition. The relief valve stack was too low to be used as a flare stack, and the base would not support a sufficiently high stack (Kletz, 1974).

A creative individual suggested that injecting steam into the relief valve discharge piping would disperse the vapour and reduce the chance of a possible unwanted fire. The suggestion was adopted and steam was routed into the 8 inch (20.3 cm) piping (Figure 2.17).

During an outage of the refrigeration unit, the relief valve opened and the steam was turned on to disperse the gases. About 12 hours later, a leak developed near the base of the tank. Steam hoses were directed near the leak to heat and disperse the vapour as the tank was emptied as sales conditions allowed. The tank could not be emptied quickly as there were no other tanks available to handle this material.

The 8 inch (20.3 cm) safety relief valve discharge line had frozen up. It is easy to understand that hot steam does not stay warm very long when it contacts −155°F (−101°C) ethylene. The steam quickly condensed and the resulting water froze. No one seemed to see the shortcomings of the original faulty suggestion.

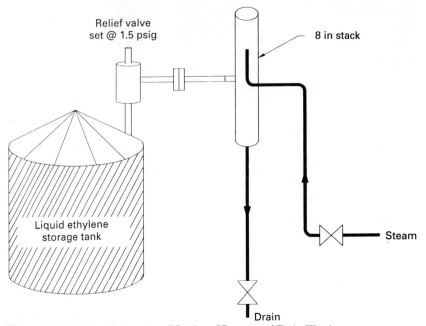

Figure 2.17 Well-intended vent modifications (Courtesy of T. A. Kletz)

Afterthoughts

Operators travelled to the area every hour to read the pressure on the tanks. The pressure gauge had a maximum pressure indication of 2 psig (13.8 kPa gauge), and it was pegged on the highest value, but no one noticed that the higher pressure or the lack of steam from the stack could signal a problem. As a matter of fact, for 11 hours before the tank cracked, operators recorded '2.0 psig' on the log sheet. No one noticed that this 2.0 psig reading was above the relief valve setting.

This incident, as well as a number of other problems with plant modifications, has been published by Kletz (1974). His book is highly recommended.

Beware of impurities, stabilizers, or substitute chemicals

Early one Sunday morning an explosion occurred in a non-flammable solvents plant. The explosion created a hole about 7 ft (2 m) in diameter and 3 ft (1 m) deep. The explosive forces were created within a temporary vessel that was 15 inches (38 cm) in diameter and 6 ft (1.8 m) high.

The top and bottom of this temporary vessel were constructed of 24 inch (61 cm) diameter, ¾ inch (1.9 cm) thick flat steel plate that was held together with 16 ½ inch (1.27 cm) bolts on each head. The explosive forces propelled the top 450 ft (140 m). The bottom plate was found in the big hole in the ground. A fire occurred and was quickly extinguished. Fortunately the only injuries were a bruised knee and an eye irritation.

On the previous Friday a tank of an organic solvent product was found to be high in acidity and was out of spec in colour. The solvent was a non-flammable cleaning fluid used to clean grease and soil from common metals. An inhibiting stabilizer (nitromethane) is added to keep the solvent from corroding metals.

It was the practice to treat slight acidity in such solvents in a neutralizer that was filled with solid caustic soda flakes or beads. Such treatment was common on several solvent intermediates and products, so it was a good intention to use caustic soda to improve this product.

However, this particular product was a solvent treated with about 2% of nitromethane stabilizer and was not like the other intermediates or products. If any chemist from the control lab had been contacted, he would have opposed this neutralization scheme.

The nitromethane in contact with the caustic soda probably created a highly unstable sodium fulminate. The explosion occurred after about 1600 gallons of the solvent had been successfully neutralized.

Afterthoughts

This accident occurred over 20 years ago, but it serves as a good example of the need to study any chemical modifications. Nearly every experienced chemical plant worker understands the problems of just a few per cent of flammables in air, or just a small amount of acid transferred into a non-resistive system, but we sometimes take the other chemicals for granted.

A gas compressor is protected from dirt, but the plant catches fire

Just before a plant start-up, the operations team became concerned that the piping upstream of a flammable gas compressor may not have been cleaned sufficiently to prevent dirt from entering the precision machinery. A temporary filter was installed in the suction line. The operations team was correct; there was a sufficient amount of foreign material in the upstream piping to plug up the filter (Kletz, 1974).

Unfortunately the temporary filter was placed between the compressor and a low suction pressure trip. The compressor suction pressure was reduced to a vacuum and air intruded into the system. The air reacted within the system. A decomposition occurred further downstream and a major fire caused many months delay in the start-up of this unit.

Afterthoughts

Just before a plant start-up there are lots of details that must be reviewed, and there are pressures to start up on the target date. Therefore it is not surprising that some modifications introduced by clever individuals during this time result in serious unforeseen consequences. This is especially the case if there is not a disciplined method to deal with changes of process technology.

The lighter side

The duck pond at a company guesthouse was overgrown with unsightly weeds. The company's water treatment specialist was approached for help. He added a herbicide to kill the weeds, but failed to realize that the herbicide also contained a wetting agent. This modification wetted the duck's feathers and the ducks sank (Kletz, 1990).

A review of good intentions

Each of the items covered as 'good intentions' showed a flaw in judgement. None of the principles were unique and no decision was so urgent that the change needed to be accomplished immediately. All technical plant employees could easily understand the problems after they occurred and most of the technical employees could have spotted the weaknesses prior to the incident, if they were afforded the opportunity for a well disciplined review (see Chapter 9).

Happily there were no significant injuries reported in these accidents that caused over one million dollars of losses. And yet, some accidents resulting from similar modifications have caused injuries and even deaths.

References

Bodurtha, F.T. (1976) Explosion hazards in pollution control. *Loss Prevention,* American Institute of Chemical Engineers, New York, **10**, 88–90

Doyle, W.H. (1976) *Loss Prevention,* American Institute of Chemical Engineers, New York, **10**, p. 76

Institution of Chemical Engineers. *Over- and Under-Pressuring of Vessels*, Hazard Workshop Module 001. Available as a training kit with thirty-four 35 mm slides, booklets, teaching guides, etc. from the Institution of Chemical Engineers, 165–171 Railway Terrace, Rugby CV21 3HQ, Warwickshire, UK

Kletz, T.A. (1974) Case histories on loss prevention. *Chemical Engineering Progress*, 70, 80. Also available as *Hazards Of Plant Modifications*, Hazard Workshop Module 002. Available as a training kit with thirty-eight 35 mm slides, booklets, teaching guides, etc. from the Institution of Chemical Engineerss, 165–171 Railway Terrace, Rugby CV21 3HQ, Warwickshire, UK

Kletz, T.A. (1990) *Critical Aspects of Safety and Loss Prevention*, Butterworths, London, p. 221

Sanders, R.E., Haines, D.L. and Wood, J.H. (1990) Stop tank abuse. *Plant/Operations Progress* (January), pp. 61–65

Chapter 3

Changes made to prepare for maintenance

Within a typical industrial chemical company, the annual cost of maintenance is 5–7% of the original assets. A number of significant maintenance activities require very little interruption of continuously operating chemical plants. These everyday activities, such as repairs to a spare pump or spare compressor after it has been properly isolated and cleared of fluids, or overhaul of a fully spared filter or painting, are accomplished on a routine basis.

However, other preparations for maintenance activity which seem innocent can jeopardize or ruin equipment and result in injury. The following examples of system modifications were made to prepare for maintenance. Each unwise modification created equipment damage. These incidents were simple changes which were made to isolate, protect, clean or clear equipment, and the methods chosen created problems. Some equipment was subjected to partial vacuum or positive pressure beyond the tolerable design limits.

A tank vent is routed to a water-filled drum to 'avoid' problems

It was necessary to perform welding near the roof on a 10 ft (3 m) diameter atmospheric tank within a cluster of 25 ft (8 m) high tanks containing flammable liquids. The conscientious foreman was concerned that a fire hazard would exist if alcohol was pumped into the tank. The foreman believed that any flammable fumes released from the goose-neck top vent constituted too much of a risk with burning and welding about to start (Institution of Chemical Engineers, Hazard Workshop Module 001).

Knowing that these alcohol fumes could be readily absorbed in water, the foreman arranged to have a hose connected to the top vent. The hose routed any fumes 'safely' to a drum of water. Figure 3.1 shows the significant points. As the welding job proceeded, operations started to pump out the tank with the 'safeguarded' vent and the tank top partially collapsed. The damage was minor.

The foreman did not appreciate the frail nature of these standard low pressure tanks. Many low-pressure tanks are designed for positive pressures of just 8 or 10 inches of water (2–2.5 kPa gauge) and just 3 or 4 inches of water (¾–1 kPa) vacuum. Using a hose to modify the system

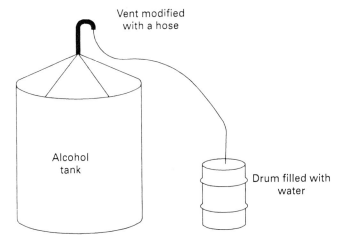

Figure 3.1 'Avoiding' problems creates problems

addressed the immediate problem of dissolving any alcohol fumes, but it created a situation in which the tank could be damaged by positive pressures and vacuum conditions well beyond its design conditions.

Afterthoughts

Typical healthy human lungs can develop 50 or more inches of water (12.5 kPa gauge) pressure. It is easy to compare the pressure that lungs can supply to the typical 10 inch water pressure (2.5 kPa gauge) positive design pressure of low-pressure storage tanks. A standard medium-sized party balloon requires about 25–30 inches of water pressure (6.3–7.5 kPa gauge) to inflate. So if you can blow up a balloon, you can damage an isolated atmospheric tank with your lungs, provided that the air does not leak out faster than you can exhale and that you are very persistent.

The Institution of Chemical Engineers (England) offers an excellent training kit which points out the tender nature of tanks. Figure 3.2 is an updated version of one of the slides within that training kit (Institution of Chemical Engineers, Hazard Workshop Module 001).

Preparing to paint large tanks

Within chemical and petrochemical plants, conservation vents and goose-neck vents protect many low-pressure tanks. Some of these tanks do not have an overflow line. There are numerous stories of well meaning painters, foremen, or field engineers damaging tanks. These individuals become concerned with possible blasting sand or paint overspray contamination of the fluids stored within the vessel when the top of the tanks are being maintained.

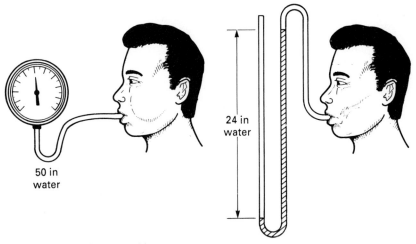

Figure 3.2 Human lung capacities

Conservation vents have been wrapped with polyethylene film to reduce the chance of contamination. In this unprotected state someone occasionally pumps out the tank or a thunderstorm brings cooling rain showers, and the resulting partial vacuum creates damage. Many plants have learned this lesson the hard way.

Preparing a brine sludge dissolving system for maintenance

A major chemical plant dissolved slurry solids in a series of three glass fibre vessels. This section of the plant was considered one of the non-hazardous areas of this chemical complex as it handled a relatively cool sodium chloride (table salt) water stream which was saturated with insoluble salts such as calcium carbonate and magnesium carbonate. The slurry contained about 30% calcium carbonate. Insoluble carbonates are treated with hydrochloric acid in two agitated vessels to allow a reaction to liberate carbon dioxide and form a soluble salt. The reaction is a simple one:

$$CaCO_3 + 2HCl \rightarrow CaCl_2 + H_2O + CO_2 \uparrow$$

The dissolving system was scheduled for maintenance while the rest of the unit continued at full production rates. The chemical process operator was concerned about the available room in the sump as the slurry accumulated during the 2–3 hour outage. He drastically increased the slurry rate into the system for about a 90 minute period before the shutdown. However, he increased the flow of acid into the system only slightly during that time. The basic process is shown in Figure 3.3.

The maintenance needs were simple. An oil change was scheduled for the no. 2 solids dissolving tank agitator gearbox and several slurry valves were to be replaced.

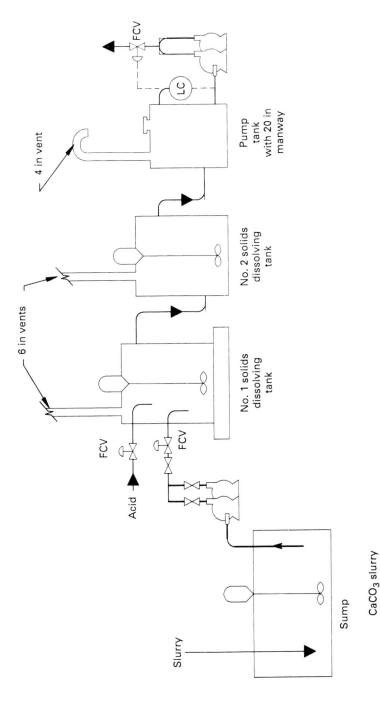

Figure 3.3 Slurry dissolving system (Courtesy of J. M. Jarnagin)

To prepare for maintenance, the slurry pumps were shut down and the no. 2 dissolving tank agitator was shut down. Erroneously, the weak acid flow rate remained constant for about 2 hours into the no. 1 dissolving tank and overflowed into the no. 2 solids dissolving tank.

After maintenance was completed (it took 2 hours) an operator restored the no. 2 solids dissolving tank agitator. Instantly, a rumbling noise was heard and a weak acid solution erupted from the no. 2 solids dissolving tank, shooting up as high as 30 ft (9 m) into the air.

This overpressure damaged two glass fibre tanks. Three cracked nozzles on the no. 2 solids dissolving tank required repairs. Damages to the pump tank were more severe. A 20 inch (50 cm) manway nozzle and its accompanying blind flange blew off and a 4 inch (10 cm) vent nozzle and stack separated from the smaller tank.

What happened in the brine system?

Hydrochloric acid was supplied by pipeline from another area of the same chemical complex. The acid was a byproduct of a process which manufactured large quantities of flashing flammable liquids. Acid samples were promptly collected and analysed for the presence of flammables, but none were found.

An investigating team reviewed strip charts of the effluent pH, acid flow rate changes and slurry flow rate changes. It was concluded that over 100 lb (45 kg) of calcium carbonate may have accumulated in the number 2 solids dissolving tank prior to shutdown. Investigators determined that the acid strength continued to increase in the system during the 2 hour outage, until the pH dropped well below 2.0 according to the effluent pH strip recorder.

When the agitator was restored, the strong acid and finely divided calcium carbonate solids instantaneously reacted. This instantaneous reaction created $400 \, ft^3$ $(11.3 \, m^3)$ of carbon dioxide within a relatively small tank which was nearly filled with salt water.

Follow-up action

The investigating team recommended that the vent stacks on each tank be increased from 6 inches (15 cm) in diameter to 10 inches (25 cm), i.e. about a 300% increase in cross-sectional venting area, but they knew it was not enough. A revision was made to the pH-monitoring system. Agitator start-up buttons were relocated from the top platform to ground level.

New procedures were instituted. Upon system shutdown, the operators are required to continue agitation and acid flow for at least 30 minutes after the slurry feed is shut down. On start-up, the agitators and acid feed will be restored before starting up the slurry pumps.

Violent eruption from a tank being prepared for repairs

A seeping leak was discovered at the bottom of a 6 m (20 ft) diameter storage containing 73% caustic soda at 121°C (250°F). The Shipping Department had intended to load a number of railway cars with the hot

73% caustic soda from the leaking low-pressure, flat-bottom, straight-side tank. Tank cars were to be loaded during the day shift, and the night shift chemical process operators were left written instructions to dilute the remaining caustic soda with hot (60°C or 140°F) water. This 73% caustic soda solution freezes easily if not diluted during a tank clearing project (Institution of Chemical Engineers, 1987). The Shipping Department loaded one less railway tank car than the operations foreman anticipated. Much more caustic soda remained in the tank than was expected by operations supervision when the instructions were written. Perhaps there were 90 tons of 73% caustic soda remaining when supervision believed only a small heel would be left.

Operations followed the instructions in the log book and started adding hot water to the vessel. It took about 11 hours to increase the tank level by 3.4 m with hot water. (Institution of Chemical Engineers, 1987).

The water addition was completed early in the morning, about 5:45 a.m. About 15 minutes later, the operator opened up the compressed air valve to the lance to 'roll' or mix the contents within the tank as the log book instructions directed. The mixing of water and strong caustic soda solutions is strongly exothermic.

Thirty minutes later, about 6:40 a.m., the tank started making a roaring sound like a jet plane. The vessel buckled upward and a mixture of steam and diluted caustic soda blew out of the large manway shooting up 25 m (80 ft) in the air. The explosive release only lasted about 30 seconds, and was followed by non-violent steaming for an additional 30 minutes.

Violent boiling spewed several tons of caustic soda droplets from an open manway on the roof of the tank. These droplets rained down in the vicinity of the tank and travelled downwind inside the plant. A light wind carried the caustic soda over railway cars on a nearby in-plant railway spur and further to employee cars in the main car park. Corrosive chemical fallout was detected over 2 km (1¼ miles) away. The plant emergency brigade used fire hoses to wash down the affected areas.

Thirty railway tank cars required a new paint job and some of the employees' cars required cosmetic repairs. Property damages for repainting rail cars and other clean up totalled over $125 000 (US, 1985). Although there were no serious injuries, an employee in the car park (US, parking lot) received enough caustic droplets in her hair to require a visit to the beauty parlour. If this incident had occurred about 30 minutes later, the day shift would have been arriving and there would have been a potential for more injuries as the car park filled with activity.

Afterthoughts

After this embarrassing and potentially dangerous incident, the following procedures were implemented:

1. Precise operating procedures will be issued for diluting caustic soda solutions.
2. Air-rolling will be simultaneously started when any water addition is required, so only a little material at a time can release the heat of solution.

An explosion while preparing to replace a valve in an ice cream plant

Food processing employment is no doubt viewed by the general public as being a 'much safer' occupation than working in a chemical plant, but in recent years the total recordable case incident rates for the food industry is about three to five times higher than that for the chemical industry, according to the US National Safety Council. However, in terms of fatal accident frequency rates, the food industry and the chemical industry have experienced similar rates in recent years (National Safety Council, 1985, 1986, 1987, 1988, 1989). This accident occurred within an ice cream manufacturing facility, but could happen within any business with a large refrigeration system.

An ice cream plant manager was killed as he prepared a refrigeration system to replace a leaking drain valve on an oil trap. The victim was a long-term employee and experienced in using the system. Evidence indicates that the manager's preparatory actions resulted in thermal or hydrostatic expansion of a liquid-full system which created pressures extreme enough to rupture an ammonia evapourator containing $5\,ft^3$ (140 litres) of ammonia (Refrigerating Engineers and Technicians Association, 1990). The local fire department's investigation revealed that when the overpressure rupture occurred, the rapid release of anhydrous ammonia along with a small quantity of lubricating oil discharged from the system was the fuel for a simultaneous combustion explosion. It was theorized that electrical arcing in the refrigeration system was the ignition source. The overpressure rupture and the combustion explosion created costly damages as high as \$250000 (US, 1980). The manager was killed; two other employees in the room were injured; mechanical systems were destroyed in the blast area; and there was damage to the front offices.

The state police report quoted a serviceman who reported that all of the valves on the evaporator were shut off and the fire chief found a steam hose steaming underneath the machine. The serviceman had previously seen steam used to melt ice buildup around the valves and to heat the oil in the machine to make it easier to drain the oil (Refrigerating Engineers and Technicians Association, 1990).

The district engineering inspector's report confirmed that the machine was completely isolated and the ammonia system was not protected by a safety valve or any type of pressure relief device. The cause of the accident was the presence of the steam hose underneath the equipment to melt the ice on the drain valve or to help evacuate the system. The inspector's report could not determine if these preparations were intentional or accidental (Refrigerating Engineers and Technicians Association, 1990).

Afterthoughts

Chemical process equipment and especially piping can be blocked in liquid with no vapour space. Many fluids, including liquid ammonia, liquid chlorine, liquid vinyl chloride, and benzene, if improperly isolated, can develop hundreds of pounds per square inch (thousands of kilopascals) of pressure with fluctuation in night-time and daytime temperatures. The

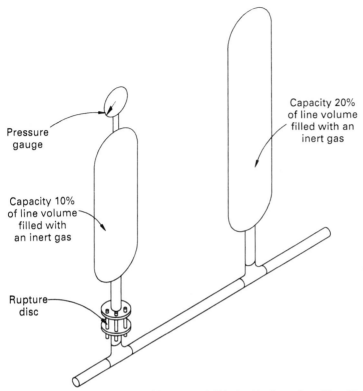

Pressure
gauge

Capacity 20%
of line volume
filled with an
inert gas

Capacity 10%
of line volume
filled with
an inert gas

Rupture
disc

Figure 3.4 Expansion chambers (Courtesy of Chlorine Institute, Inc., New York)

liquid tries to expand as a result of its physical properties, but it is prevented by confinement.

Procedures to ensure proper isolation of such fluids in pipelines should be put into use and must be backed up with the proper overpressure device. If the trapped fluid is highly flammable, has a high toxicity, or is otherwise very noxious and not a candidate for a standard rupture disc or safety relief valve, the designers should consider an expansion bottle system like that which the Chlorine Institute recommends to prevent piping damage (Figures 3.4 and 3.5).

A chemical cleaning operation kills sparrows, but improves procedures

Various methods of cleaning vessels, other equipment and pipelines are regularly employed as a preparation for maintenance or the actual maintenance itself. Various acids and alkalis are used for this activity. There can be risks associated with such reactive chemicals, as noted in the following paragraphs.

Figure 3.5 An engineer observes the pressure gauges on chlorine expansion bottles

When a number of dead sparrows had accumulated around the polymerization unit, refinery management realized it needed to develop more rigid policies governing chemical cleaning of equipment. A chemical cleaning business had been hired to acid clean a caustic scrubbing system of the catalytic polymerization unit. The chemical reactions were not considered. No one predicted that the acid would react to form hydrogen sulphide and a heavy concentration of the gas was released to the atmosphere. No personnel were injured, but a number of birds were not that fortunate (American Petroleum Institute, 1981).

After the incident, it was determined that the refinery needed to more closely plan and supervise such chemical cleaning operations. A planning group was composed of one supervisor each from the operations, maintenance, inspection, technical, and safety departments.

1. The group should decide how to vent the system during the cleaning operations, such as to the flare, through a caustic wash, or directly to the atmosphere.
2. The technical department was charged with the responsibility to prepare sketches of the equipment and decide on the details such as location of the connections, a step-by-step procedure for cleaning, safe disposal of the cleaning fluid, personal protective equipment required, etc.
3. Copies of sketches and any other pertinent instructions would be issued to the contractor doing the work, and to the operating, maintenance and inspection departments.
4. Safe work permits would be issued by the operating foreman. The line-up would be checked by the area supervisor or the shift supervisor.

This approach to form a team of specialists to evaluate a proposal to chemically clean refinery equipment is a concept that will be discussed in Chapter 9.

Other cleaning, washing, steaming and purging operations

Cleaning, washing, steaming-out and purging of equipment, vessels and pipelines in preparation for maintenance can create problems. Often equipment cleaning does not attract much attention and is frequently carried out by individuals with the least amount of technical training. A few problems that have been reported will follow.

A 30 000 gallon (115 m^3) atmospheric tank in a fine chemical plant in West Virginia was being readied for a different batch. As part of the changeover in products, the tank had to be cleaned out. Steam cleaning was the usual process and no one seemed to consider the brutally cold weather which was nearing 0°F (-18°C).

The fine chemical tank was steamed out for a considerable time and the steam valve was shut. Unexpectedly, the tank partially collapsed from the rapidly condensing steam, despite the fact that an 18 inch (46 cm) manway was completely open.

In a different incident, a tank containing a solution of ammonia and water was being emptied and cleaned out. After it was emptied of liquid ammonia, water was added to rinse the vessel. While the water was being added, the sides of the tank were sucked in. The ammonia vapour remaining in the tank dissolved in the water so rapidly that air could not enter through the vent to prevent the collapse (T.A. Kletz, personal communication).

Calculations were made. A vent of about 30 inches (76 cm) in diameter would have been needed to prevent a vacuum from being formed during the ammonia absorption incident.

Every plant has a story of well-meaning plant employees washing a pipeline or vessel of an acidic or reactive material that went sour. Improperly washing a carbon steel vessel in strong (about 98%) sulphuric acid has left wall thickness damage. Underestimating the heat of solution or heat of reaction of water washing out other chemicals has resulted in clouds of toxic fumes.

A tragedy when preparing for valve maintenance

A catastrophic explosion and major fire occurred within a major refinery as a system was being prepared for valve maintenance. A flashing flammable fluid, isobutane, with a boiling point of 11°F (−12°C), was stored in two spherical tanks. The spheres were connected to an alkylation unit via a 10 inch (25 cm) line. The line pressure was about 50 psig (345 kPa gauge) and one of the valves in this underground system was in an open pit (Vervalin, 1973).

A refinery operator saw bubbles in the water that covered the 10 inch (25 cm) valve in the pit. The employee proceeded to try to clear the underground pipeline by flushing the line with a 110 psig (760 kPa gauge) water supply from the unit sphere no. 1. When it was determined that water had flushed the flammables from the pipeline and into the sphere no. 1, the valve to sphere no. 1 was closed and the operator started to open into the sphere no. 2.

High water system pressure of nearly 110 psig (760 kPa gauge) against the 10 inch closed valve to sphere no. 1 forced the bonnet to fly off the weakened valve in the pit and a geyser of water was seen. The operator saw the geyser, sensed something was wrong and mistakenly closed the valve to the no. 2 sphere and reopened the valve to the no. 1 sphere (Vervalin, 1973).

This action allowed an estimated 500 barrels of isobutane to escape and ignite explosively. The resulting fire spread to other equipment and burned for two weeks. It was later determined that the 10 inch valve in the pit had badly corroded bonnet bolts because of a leaky sulphuric acid line that ran nearby (Vervalin, 1973).

One company report indicated that the explosion occurred about 4:45 a.m. and residents of the area were awakened by a frightening roar to see smoke and flame. Calculations indicated that the vapour cloud explosion had the force of 10–12 tons of TNT. There were seven deaths, 13 injuries and an insured loss exceeding $35 million (US, 1967) (Goforth, 1969).

Afterthoughts on piping systems

Corrosion is a serious problem throughout the world and external corrosion can often be observed on piping, valves and vessels within chemical plants. Each plant must train its personnel to observe serious corrosion and external chemical attack.

Piping is often not appreciated as well as it should be. As many chemical plants are beginning to grow older, more piping corrosion problems will occur. It is important that piping is regularly inspected so that plant personnel are not surprised by leaks and releases. The American Petroleum Institute (API) understands this need for piping inspection and has recently issued API 574 (American Petroleum Institute, 1990). This describes piping standards and tolerances, as well as offering practical basic descriptions of valves and fittings, and includes 16 pages covering inspection. Various topics including reasons for inspection, frequency of inspection, inspection tools and inspection procedures are included.

Figure 3.6 Pipe hangars corrode and pipe drops to the ground

An inspection procedures section details the areas of piping which are the most subject to corrosion, erosion and other forms of deterioration. Inspection of piping systems while in service is addressed, including examining flanged joints, packing glands, expansion joints, and various supports, including pipe shoes, hangers and braces. Various non-destructive test methods such as inspection for external corrosion, ultrasonic thickness inspection and radiographic inspections are also addressed.

A review of changes made to prepare for maintenance

Changes made for maintenance can sometimes require non-routine activity. Several previous case histories offered approaches and procedures to avoid mistakes which can be disruptive or unsafe. Procedures must be developed and reviewed by sufficiently trained individuals.

The US Occupational Safety and Health Administration, OSHA, has issued a comprehensive law entitled 'Process Safety Management of Highly Hazardous Chemicals.'

A section of the appendix to that law reads:

Nonroutine Work Authorizations. Nonroutine work which is conducted in process areas needs to be controlled by the employer in a consistent manner. The hazards identified involving the work that is to be accomplished must be communicated to those doing the work, but also to those operating personnel whose work could affect the safety of the process. A work authorization notice or permit must have a procedure that describes the steps the maintenance supervisor, contractor representative or other person needs to follow to obtain the necessary clearance to get the job started . . .

These are good words, but may be easier said than done.

References

American Petroleum Institute (1981) *Safety Digest of Lessons Learned*, Section 4, API, Washington, D.C., p. 49

American Petroleum Institute (1990) *Inspection of Piping, Tubing, Valves and Fittings*, Recommended Practice 574, 1st edn, API, Washington D.C.

Goforth, C.P. (1969) Functions of a loss control program. *Loss Prevention*, American Institute of Chemical Engineers, New York, **4**, p. 1

Institution of Chemical Engineers. *Over- and Under-Pressuring of Vessels*, Hazard Workshop Module 001. Available as a training kit with thirty-four 35 mm slides, booklets, a teaching guide, etc. from the Institution of Chemical Engineers, 165–171 Railway Terrace, Rugby CV21 3HQ, Warwickshire, UK

Institution of Chemical Engineers (1987) *Loss Prevention Bulletin, Articles and Case Histories from Process Industries throughout the World*, **078**, pp. 26–27

National Safety Council (1985) *Work, Injury and Illness Rates 1985*, NSC, Chicago, IL

National Safety Council (1986) *Work, Injury and Illness Rates 1986*, NSC, Chicago, IL

National Safety Council (1987) *Work, Injury and Illness Rates 1987*, NSC, Chicago, IL

National Safety Council (1988) *Work, Injury and Illness Rates 1988*, NSC, Chicago, IL

National Safety Council (1989) *Work, Injury and Illness Rates 1989*, NSC, Chicago, IL

OSHA Part 1910.119 (1992) Process Safety Management of Highly Hazardous Chemicals: Explosives and Blasting Agents: Final Rule, Appendix C, US Department of Labor, Occupational Safety and Health Standards, Federal Register, Washington DC, Feb 24, p. 6414

Refrigerating Engineers and Technicians Association (1990) One man's death: an investigative report. *The Technical Report*, **3**, No. 2

Vervalin, C.H. (ed.) (1973) *Fire Protection Manual – For Hydrocarbon Processing Plants*, 2nd edn, Gulf Publishing Company, Houston, TX, p. 90

Chapter 4

Modifications introduced in maintenance activity

When a single piece of equipment or an entire unit is shut down for maintenance or if a maintenance crew is working on a system, it is everyone's expectation that the system will be improved and made more reliable. But occasionally, these maintenance activities inadvertently introduce an unwanted modification into a carefully engineered plant. In many industrial chemical plants the number of individuals involved in maintenance activity is about a third of the entire plant's employees, so it is logical to think that this particular activity has a significant role in plant accidents due to modifications. Trevor Kletz often stated during his lectures that, in his industrial experience, the preparation for maintenance and the actual maintenance activity accounted for more chemical plant accidents than any other single cause.

Maintenance activity is also drawing the attention of US regulators. The Occupational Safety and Health Agency (OSHA) has proposed OSHA Regulation 1910.119. It requires facilities which handle highly hazardous chemicals to implement a 14-point process safety management programme. Three of the elements 'Hot work permits,' 'Mechanical integrity,' and 'Contractor safety', impact directly on a maintenance programme.

When performing turnarounds, repair, or preventive maintenance, it is necessary for the job to be well planned, with approved procedures, proper materials and qualified repairmen. Good communications at the job site are imperative.

One simple example of poor communication in maintenance occurred after the plant staff marked the location of an underground electrical bus with a series of flags on the surface. No one discussed the job with the backhoe operator, so when he saw the flags, the hoe operator assumed that the flags indicated the location of the path to be dug. So this well meaning backhoe operator severed the sub-surface electrical cables and a large portion of the chemical plant facility was blacked out. The fate of the backhoe operator was not mentioned (Lorenzo, 1990).

Reboiler passes the hydrotest and later creates a fire

A flowing combustible heat transfer oil was routed through the shells of four heat exchangers after it was heated within the coils of a furnace

41

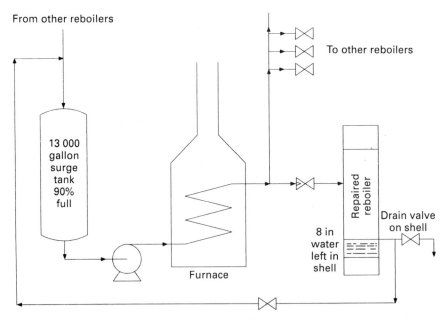

Figure 4.1 Water vaporizes and ruptures a vessel (Courtesy of T. A. Kletz)

(Figure 4.1). One of the heat exchangers required considerable repairs and was taken out of service. While repairs were underway on one exchanger, hot circulating oil at about 460°F (240°C) was still routed through the other three exchangers (T.A. Kletz, personal communication).

At the conclusion of the repairs, it was decided to hydrostatically test the unit while in the normal vertical position. The test was conducted to ensure mechanical integrity and to check the heater for leaks before start-up. When no leaks were found, the water was drained from the shell side. Unfortunately, the lower drain valve nozzle was not located on the very bottom of the unit and it allowed about 21 cm or about 8 inches of water to be trapped.

When this fourth exchanger was returned to service, hot oil swept the trapped volume of water into a surge tank where it explosively vaporized. The surge tank roof was blown off in one piece; the bursting pressure was 450 psig (3100 kPa gauge). This explosion created a fine mist-like cloud which ignited immediately and formed a fireball 120 m (400 ft) in diameter (T.A. Kletz, personal communication).

Many individuals do not realize the tremendous increase in volume that occurs when water is vaporized to steam. Hazard Workshop Module 1 developed by the Institution of Chemical Engineers (Rugby, UK) contains a slide which very simply illustrates this huge increase in volume (Figure 4.2).

WATER EXPANDS IN VOLUME
ABOUT 1600 TIMES UPON
VAPORISATION AT 100°C
AT ATMOSPHERIC PRESSURE!
+100°C

I PINT + 100°C = (5×40 GAL DRUMS)

Figure 4.2 Man with a pitcher of beer (Courtesy of the Institution of Chemical Engineers)

A tank explodes during welding repairs after passing a flammable gas test

No one wants to modify a tank with an accidental explosion; but here are two examples. Kletz (1984) lists Myth No. 9, 'If a flammable has a high flash point it is safe and will not explode.' This small booklet lists a number of statements that chemical plant employees have accepted, often without being sceptical of the limitations of these myths.

These myths are usually not completely erroneous and often contain a measure of truth. However, they are often believed to be absolutely true, despite the doubtful accuracy. Another of the 44 Myths listed in the booklet is: 'It must be safe, as we have done it this way for years without an accident.'

Of course, handling a high flash point material can be considerably safer than handling a low flash point liquid if, and only if, the high flash point material is maintained well below its flash point. But once a high flash point fluid such as fuel oil, heat transfer oil, or other high boiling combustible liquid is heated above its flash point it becomes equally as dangerous as petrol or flammable gases.

Additional hazardous conditions can occur with aerosols or fine mists of combustible high flash point material. These mists can explode, just as combustible dusts will explode, even if the fluids in the mists are at temperatures well below their flash point (Kletz, 1984).

Kletz (1990) defines the problems of working on tanks containing high flash point oils. His direct quote introduces the next two incidents.

'Many fires and explosions have occurred when welding has been carried out on *tanks* which have contained heavy oils (or materials which polymerize to form heavy oils). It is almost impossible to remove all traces by cleaning and the welding vaporizes the oil and then ignites it. If welding has to be carried out on such tanks they should be inerted with *nitrogen* or with fire fighting foam made with nitrogen (but not foam made with air). The volume to be inerted can be reduced by filling the tank with water up to the level at which welding has to be carried out.'

A phenol tank's roof lifts as repairs are made

Repairs to the roof of a small phenol storage tank were necessary. Supervisors wanted to complete the repairs without the fuss and bother of emptying and washing the vessel if that could be avoided. This portion of the chemical plant had experienced a minor incident with hot phenol a few years before, so it was important for them to empty the tank as far as practical. It was decided to shut off the steam to the heating coil well in advance. The loss of steam allowed the heel of the tank to freeze. The melting point of this combustible material was 41°C (106°F).

The welder was directed to weld a patch-plate on the roof of the low-pressure, vertical tank. Heat from the welder's torch vaporized and ignited some phenol which had accumulated on the underside of the roof and a mild explosion occurred. The overpressure tore some of the roof seam, and, fortunately, the alert welder saw evidence of an internal fire and ran to safety just before the explosion occurred (Kletz, 1984).

Another country, another decade and a similar tank explosion

In June 1988 there was a fire and an overpressurization of a tank containing about a 3 ft (1 m) level of a combustible liquid in a tank that was 25 ft (7.6 m) high and 20 ft in diameter (6.1 m). At the time of the incident, 1:15 p.m., the tank was being repaired by a welder. Regrettably, three employees were injured and received relatively minor burns, abrasions and lacerations.

Two days before the incident, the maintenance foreman, the shift foreman, the operations engineer and two pipefitters discussed the job. The scope of the job included repairs to handrails and walkways. It was decided that burning and welding would be permitted on the structural steel that was located on top of the tank, *but no burning and welding* on the actual surface of the tank top was to be allowed.

The first day progressed with no problems, and it rained the next day, which prevented further work on the tank. On the third day, the job resumed with the same two pipefitters and a welder was added to the crew. Flammable gas testing of the area was performed on the exterior of the tank and the probe was lowered a few feet into the tank through an atmospheric vent. There was no evidence of a flammable mixture.

No one told the welder that burning and welding were prohibited on the surface of the tank's top and no one reminded the pipefitters on this day. While the job was in progress, it seemed expedient to weld a horizontal support to the top of the vessel for a section of walkway. As the welder attempted to attach the post to the thin tank top his weld 'blew through' the tank top.

It was speculated that when the welder's arc penetrated the surface of the tank top, combustible organics (a benzene derivative) ignited. Most likely, combustible material had vaporized from liquid over a period of time, condensed and solidified on the underside of the tank roof. Furthermore, it is believed that the burning organic fell through the vapour space until it ignited a dense layer of flammable vapour somewhere near the liquid surface.

The fuel/air conditions must have been ideal because fire violently created combustion gases at a rate which quickly exceeded the tank's venting capability. In less than 5 seconds, as the top ruptured, alert employees hurried off the tank. The last of the three fleeing employees recalled being propelled into the other two escaping employees ahead of him when the tank top blew open.

An accident investigation was promptly initiated. All the employees interviewed by the investigating team were generally of the opinion that the organic was practically not ignitable. Employees were familiar with handling this organic compound well below its 148°F (65°C) flash point. Perhaps they believed in the myth 'If a combustible liquid has a high flash point it is safe'.

Filter cartridges are replaced and an iron-in-chlorine fire develops

An astonished operations crew witnessed the body of a carbon steel filter spontaneously ignite and burn just after a set of glass fibre filter cartridges were replaced. This 18 inch (45 cm) diameter and 4 ft (1.2 m) high filter had been designed to remove fine particles from a heated gaseous chlorine stream. The cartridges selected to replace the spent units were in warehouse store stock but were originally specified for other chemical services within the complex. These cartridges were made with tin-coated steel cores.

As the filter with the new cartridges was returned to service, the tin spontaneously caught fire within the hot chlorine atmosphere and burned with sufficient heat to initiate an iron-in-chlorine fire. The upstream chlorine valves were blocked and the fire continued to burn until all of the chlorine gas was consumed by the fire. The steel shell of the filter and several feet of downstream piping were destroyed.

Repairs to a pipeline result in another iron-in-chlorine fire

When a welder finished welding a short section of 2 inch (5 cm) piping in a rather long insulated gaseous chlorine-equalizing line it was late on the day shift. The welds were located within about 6 inches (15 cm) of the adjacent thermal insulation that encircled the chlorine line.

When the evening shift loader came on shift, his first task was to return the repaired line to service. He pressured up the section of piping with 100 psig (690 kPa gauge) compressed air and determined that the repairs were leak free. He next opened a valve connecting the 2 inch (5 cm) pipeline to 160 psig (1100 kPa gauge) compressed chlorine gas.

Within seconds, the iron caught fire and the escaping gases roared like a jet, as a brownish-orange plume of ferric chloride drifted towards the fence line from the blazing piping. The shift loader quickly isolated the piping system by closing the block valve he had just opened, and with the oxidant source missing, the fire self-extinguished. Eye witnesses said that sparking and fire resembled a traditional Roman candle, shooting sparks of burning steel. Although the fire burnt out quickly, it was remembered for years.

Many of the supervisors and loaders were well aware that if chlorine comes in contact with iron that has been heated above 483°F (250°C) a spontaneous iron-in-chlorine fire can occur. But everyone expected that the new weld on the pipeline had cooled to well below the autoignition temperature of iron in chlorine, and it probably had. The investigation determined that the fire did not initiate at the new weld, but started under the adjacent insulation which trapped heat and did not allow the piping to quickly cool.

Repair activity to a piping spool result in a massive leak from a sphere

A section of 75 mm (3 inch) piping from a sphere was scheduled for replacement due to external corrosion at a petrochemical plant in Altona, Australia, in February 1982. The section of pipe requiring replacement was just below a single air operated isolation valve (Institution of Chemical Engineers, Hazard Workshop Module 006).

The work was considered normal maintenance and presented no special hazards. A prefabricated section of replacement piping was carried to the sphere containing hundreds of tons of a flashing flammable liquid. The boiling point of this material was 8°F (−13°C) and it was stored under several atmospheres of pressure (Institution of Chemical Engineers, Hazard Workshop Module 006).

This section of piping not only served as an outlet for the flashing flammable liquid, but it also supported the valve actuator and was connected by a lever to the valve. Brackets connecting the lever to the corroded section of piping were removed and then the piping was removed. The lever and the actuator, with a combined weight of about 20 kg (45 lbs), just hung down below the sphere and were suspended from the valve. (See Figure 4.3.)

The prefabricated piping spool was not exactly the right size, so it was being removed and returned to the workshop to make it useful. Before returning to the shop with the ill-fitting spool, the fitter and two assistants decided to move the old pipe further away. Unexpectedly, somehow, the hanging actuator suddenly dragged the valve handle to the open position.

For the next 5 hours, 670 tons (608 000 kg) of vinyl chloride escaped from the sphere which was under about 30 psig (2.4 bar) pressure. The discharging liquid jetted out self-refrigerated and froze the water that was being poured over the sphere. It was later determined that liquid vinyl chloride escaped at about 3500 litres/min (725 gal/min). (See Figure 4.4.)

The breeze held steady from the south but the speeds varied from 8 to 14 knots (16–26 km/h). Potential electrical ignition sources had been shut down and fortunately there was no ignition (Institution of Chemical Engineers, Hazard Workshop Module 006).

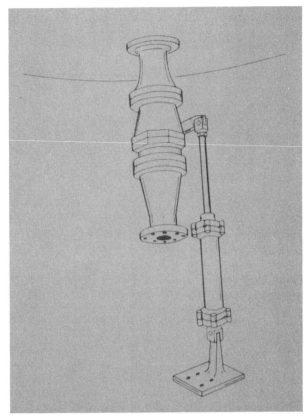

Figure 4.3 A sketch of the piping arrangement being repaired, below the sphere (From Institution of Chemical Engineers, Hazard Workshop Module No. 6)

It is ironic; the piping spool was being replaced to avoid the possibility of a leak and yet the improper maintenance modification enabled 670 tons of a flashing flammable material to escape.

The Phillips 66 incident

The tragic Phillips incident is another one in which maintenance was being performed on piping systems just below the last block valve holding back a massive amount of flammables. On 23 October 1989, 85 000 lbs (39 000 kg) of flammable gases escaped (in less than 2 minutes) from the reactor loop, which was operating under high pressure. The cloud exploded with the force of 2.4 tons of TNT according to the report by the US Department of Labor, Occupational Safety and Health Administration (OSHA). Two isobutane storage tanks exploded 10–15 minutes later and each explosion contributed to a chain reaction of explosions (Occupational Safety and Health Administration, 1990).

This tragic explosion and fire resulted in the loss of 23 employees and the destruction of nearly $750 million in property. This is the costliest single

Figure 4.4 The sphere as it discharged to the ground (Courtesy of the Institution of Chemical Engineers)

owner property damage loss to occur in the petrochemical industry. Debris from this explosion was found about 6 miles from the plant. Structural beams were twisted like pretzels from the heat of such an intense fire, and two polyethylene production plants covering about 16 acres were destroyed. (See Figure 4.5.)

High-density polyethylene reactors are long and tubular, operated under elevated pressure and temperature. The tall vertical reactors were in excess of 150 ft (46 m) high and had settling legs that were 8 inches (20 cm) in diameter. The reactor was operating at about 600 psig (4100 kPa gauge) and near 225°F (106°C).

Ethylene is dissolved in isobutane and reacts with itself to form polyethylene particles that gradually settle out of solution and are collected in one of six settling legs. Particles pass through a main 8 inch (21 cm) block

Figure 4.5 Aerial view of the destruction at Phillips 66 (Courtesy of Factory Mutual Research Corporation)

valve called a DEMCO valve to collect in a chamber. The particles are periodically isolated from the reactor, the volatiles are removed and the polyethylene is routed to other equipment for further processing. See Figure 4.6, which appeared in OSHA Reports and has been reproduced with the permission of OSHA.

According to the OSHA investigation report, the reactor was undergoing a regular maintenance procedure for the removal of a solidified polyethylene blockage on three of the six reactor settling legs. The procedure called for the isolation of the 8-inch valve and the removal of the compressed air hoses which are used to operate the valve (Occupational Safety and Health Administration, 1990).

The maintenance crew had partially disassembled the leg and managed to extract a polyethylene 'log' from one section of the leg. However, part of the 'log' remained lodged in the pipe below the 8 inch isolation valve. Witnesses stated that a mechanic went to the control room to request help from an operator. A short time later, vapour was seen escaping from the disassembled settling leg (Occupational Safety and Health Administration, 1990).

After the explosion, investigators confirmed that the 8-inch DEMCO block valve was open at the time of the accident. Furthermore, tests showed that the compressed air hoses were improperly connected in a reversed position. This improper modification during maintenance activity allowed the DEMCO isolation valve to open when the actuator switch was in the closed position (Occupational Safety and Health Administration, 1990). Over time, somehow, the company's maintenance procedures had been modified and this tragedy resulted.

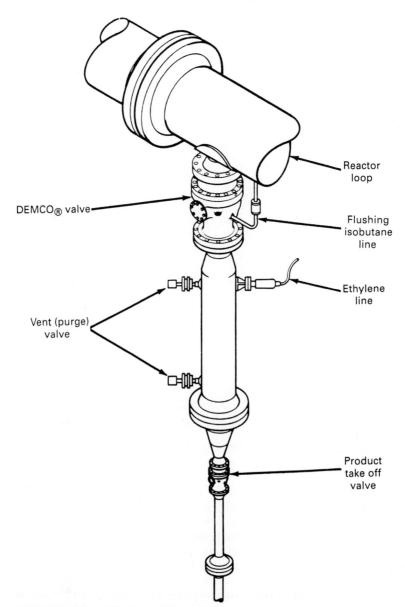

DEMCO® valve

Reactor loop

Flushing isobutane line

Ethylene line

Vent (purge) valve

Product take off valve

Figure 4.6 Typical piping settling arrangement (Courtesy of the US Department of Labor, Occupational Safety and Health Administration)

A breathing air system on a compressed air main is repaired

Before the 1970s a number of chemical plants allowed compressed air from the in-plant utility air network to be used as breathing air for individuals who were required to sandblast or enter vessels. In one such network,

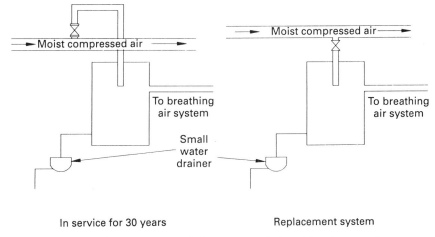

In service for 30 years   Replacement system

Figure 4.7 Modifications to compressed air systems at the breathing air connection

there were water traps installed to remove trace accumulations of water in the system just upstream of the hook-up for breathing air (Figure 4.7) (Kletz, 1988a).

The system was used for several decades without any complaints, and then a worker got a face full of water while using an air line mask within a vessel which had an unsuitable atmosphere. His companion worker who was assigned to watch him from the manway observed that something was wrong and the worker was rescued (Kletz, 1988a).

It was determined that when the compressed air main was renewed, the breathing air apparatus water trap had been piped to the bottom of the main, instead of being connected to the top as it had been for years. When a significant slug of water entered the main, the catchpot drainer was not capable of draining the water fast enough. No one realized the importance of routing the catchpot from the top of the line (Kletz, 1988a). This type of modification is difficult to detect.

A hidden blind surprises the operators

A medium-sized glass fibre acid tank experienced a sudden failure at the base just after it was returned to service. This 12 ft (3.7 m) diameter and 24 ft (7.3 m) high tank was equipped with a fill line, an overflow line, a vent line and a vacuum relief device. The vent line was routed to a scrubber. The overflow line was routed to a chemical collection sewer.

The glass fibre acid tank had just been washed out and the adjacent chemical sewer was scheduled for maintenance. In order to better protect the maintenance crew assigned to make repairs to the nearby chemical collection sewer, it was decided to blind the tank overflow line and instruct the operators to limit the liquid level to a well-defined maximum.

When isolating a hydrochloric acid tank, it is common practice, and the preferred practice at this plant, to use a two-piece blind. A sheet of Teflon® is placed toward the acid supply for corrosion resistance and this is backed up with a steel blind for strength.

The tank was returned to service. As production was routed into the vessel, the level indicator was showing a rapidly rising level. Within a short time, the tank blew apart at the base. The tank manufacturer estimated that a failure of this nature could be reached at pressures as low as 2.5 psig (17 kPa). (See Figures 4.8, 4.9 and 4.10.)

Since the ambient temperature was just below freezing, it was very easy to conclude that water froze in the small scrubber sealing the vent. It took just a few minutes to reach that conclusion, but later it was determined that other problems caused the failure.

After further review, it was determined that a Teflon® blind without the identifying protruding tab had remained in the vent line. Evidence of

Figure 4.8 A large acid tank fails due to a blocked vent and a blocked overflow line

Figure 4.9 An aerial view of the tank failure

adhesive material on the Teflon® disc suggests that it had been glued to a steel blind. No doubt, the steel blind was removed after the vessel was totally blinded off for an internal inspection. This hidden Teflon® sheet must have been blocking the vent for over a year. After the precautionary blind was installed in the overflow line, and with the hidden blind in the vent, the inerts had no place to go and the tank was pressurized to destruction.

Afterthoughts on the use of blinds

Blinds are frequently used in maintenance and the blinding method is usually customized for each type job. If a vessel or a system requires significant blinding there should be a blinding list which is used as a checklist to install and remove them. When a Teflon® blind is used with a steel blind, the Teflon® sheet should be designed and cut with a protruding

Figure 4.10 A close-up of the failed tank base

tab or 'ear'. It is obvious from this example that a circular blind without a tab can be inadvertently left behind as blinds are removed.

Other incidents in which failure to remove blinds created troubles

The water jacket on a small vessel in a soap plant was designed for a pressure rating lower than the cooling water supply pressure. Simple piping design helped to ensure that this 6 ft (1.8 m) diameter and 6 ft (1.8 m) high cylindrical vessel with a bottom dished head could not be over pressure during operation. The cooling water discharge line was engineered without a discharge valve (Griffin, 1989).

It became necessary to hydrostatically test the vessel jacket, so it was properly blinded and passed the test. The small flat-top vessel collapsed during start-up as the inlet cooling water valve was opened. Repairs were expensive and the start-up was delayed by several weeks because no one remembered to remove the blind on the 'valve-free' water discharge line (Griffin, 1989).

In another case, the scrubber which served a storage tank had been isolated to make repairs to the scrubber. Again by an oversight the blind isolating the vent scrubber from the tank was not removed when the scrubber repairs were completed (Griffin, 1989).

The glass fibre storage tank was overpressured as a transfer was being made into it. An alert operator noticed that a remote level indicator seemed to be misbehaving. He wisely stopped the unloading operation and investigated only to find that two anchor bolts were pulled out. The pressure bowed the tank's flat bottom up by several inches (Griffin, 1989).

Poor judgement by mechanics allowed a bad steam leak to result in a minor explosion

Many chemical plants rely on specialty leak repair contractors to stop steam and other utility fluid leaks in today's world. However, in the mid-1970s these contractors were not immediately available to many plant sites, so the plants had to create their own clamps or other containment systems, if shutdowns were to be avoided.

It was late on a Friday afternoon when a maintenance crew of a welder and pipefitter were assigned to stop a bad steam leak on the bonnet gasket of a 3 inch (7.5 cm) steam valve. The mechanics were instructed to fabricate a clamp which could be tightened around the valve bonnet to silence the leak of 260 psig (1800 kPa gauge) steam and condensate (Kletz, 1988b).

The leak repair clamp was a good idea, but it was one of those jobs that was easier said than done. There was minimal supervision when the crew returned on Saturday and took it upon themselves to build a steel box around the leaking valve.

The mechanics designed a temporary box that lacked most engineering design considerations as it was made of flat plate 3 ft (0.9 m) high, 2 ft (0.6 m) wide and a little over 14 inches (0.36 m) deep. The box was fabricated of ¼ inch (0.64 cm) steel plate and included a 2 inch (5 cm) vent pipe to relieve the steam as the box was encasing the valve. Even the 150 psi rated flange on the 2 inch piping was unsuitable for the operating pressure and temperature.

Upon completion of the box on Saturday, a self-assured and rather cavalier maintenance employee closed the valve on the relief line. The box started to swell. The employee quickly opened the valve and told the shift supervisor that they needed to return to finish the job on Sunday.

The employees returned on Sunday to complete the job. They decided to strengthen this poorly constructed box by welding a 2 inch (5 cm) angle iron stiffener around its middle. After the angle iron 'improvement' was completed, the mechanic closed the valve and moved away. It was a good thing he did move away rapidly because the box quickly exploded (Figure 4.11).

Figure 4.11 A 'repair' box which was improperly used to stop a steam leak

This was a chemical plant which only purchased high quality ASME coded vessels and had a staff of engineers that could easily calculate maximum allowable design pressures. Chemical plant supervision must be clear on instructions and be vigilant about craftsmen changing the scope of the job, especially during the 'off shifts', when supervision may be very limited.

After the explosion, engineers and inspectors examined the box. Calculations indicated that the box could hardly be expected to contain 50 psig (340 kPa gauge) because of its square construction and was totally incapable of containing the 260 psig (1800 kPa gauge) steam pressure from the leaking valve bonnet. Furthermore, the welds were not full penetration and the flanges and valve were also of insufficient design strength (Kletz, 1988b).

This incident illustrates that we must train our mechanics to seek approval of any changes in job scope before they independently initiate alternative approaches. We cannot assume that because we employ qualified craftsmen and graduate engineers, they will never carry out repairs with foolish or improper methods. Chemical plants must be continually vigilant for unrequested modifications.

The Flixborough disaster and the lessons we should never forget

The largest single loss by fire or explosion ever experienced in the UK occurred on 1 June 1974, as a result of an unwise plant maintenance modification. A total of 28 individuals lost their lives, 36 more employees experienced injuries, and 53 more non-employees had recorded injuries. The estimated damage was about $63 million (US, 1976) (Warner, 1975).

On the banks of the Trent river this plant produced caprolactam, a basic raw material for making Nylon 6. The process area consisted of six reactors in which heated and pressurized cyclohexane was oxidized by introducing air and catalyst. The reactors were about 12 ft (3 m) in diameter and about 16 ft (5.3 m) high. The reaction only converted about 6% of the cyclohexane, while 94% of the flammables were recirculated, resulting in large inventories relative to the production rates (Warner, 1975).

The six reactors operated at about 8.8 kg/cm² (125 psig) and at a temperature of about 155°C (310°F). The normal boiling point of cyclohexane is 81°C (178°F); therefore any leaks of 155°C material would instantly form a vapour cloud. Each reactor was installed about 14 inches (36 cm) lower than its predecessor and the circulating liquid flowed by gravity from one unit to the next. The six reactors were connected by short lengths of 28 inch (61 cm) diameter stainless steel pipes and stainless steel expansion joints. These pipes were only partially liquid-filled and the off-gases flowed to equalize the pressure (Fire Protection Association, 1975).

A crack about 6 ft (2 m) long was discovered on the no. 5 reactor and the plant was shut down on 27 March 1974. It was decided to remove the reactor and to construct a bypass pipe to allow the plant to stay in operation using the remaining five reactors. This bypass piping would route cyclohexane from reactor no. 4 to reactor no. 6.

The plant mechanical engineer had previously left the company and there was no mechanical engineer on the site. No one involved with this bypass piping project appeared to appreciate that connecting reactor no. 4 with reactor no. 6 was anything but a plumbing job. No 28 inch (61 cm) piping was available on site, but there was some 20 inch (51 cm) piping at hand. The 20 inch piping was fabricated with two mitres and installed between two 28 inch bellows (generally called 'expansion joints' in the United States). This bypass piping was supported by scaffolding normally constructed for workmen working at elevated positions (Fire Protection Association, 1975).

The 'Minutes of Proceedings of the Court of Inquiry Into the Disaster which occurred at Nypro (UK) Ltd Flixborough' stated in part (Fire Protection Association, 1975):

'No one appreciated that the pressurized assembly would be subject to a turning movement imposing lateral (shear) forces on the bellows for which they are not designed. Nor did anyone appreciate that the hydraulic thrust on the bellows (some 38 tonnes at working pressure) would tend to make the pipe buckle at the mitre joints. No calculations were done to ascertain whether the bellows or pipe would withstand these strains; no reference was made to relevant British Standard or any other accepted standard; no drawing of the pipe was made to the designer's guide issued by the manufacturers of the bellows; no drawing of the pipe was made, other than in chalk on the workshop floor; no pressure testing of the pipe or the complete assembly was made before it was fitted.'

The report further commented on the poor support for the piping. Nevertheless, the plant restarted on 1 April 1974, in this configuration, and operated successfully until late May 1974. In order to repair a leaking sight glass, the plant shut down on 29 May 1974. The plant was restarted, with steam heating beginning about 4:00 a.m. and the explosion about 5:00 p.m. The pressure wave from the explosion and ensuing fire involved 433 000 gallons (1 640 000 litres) of flammable liquids and damaged structures up to 8 miles (13 km) away.

The main recommendation of the Official Inquiry was *'Any modification should be designed, constructed, tested, and maintained to the same standards as the original plant.'*

Do piping systems contribute to major accidents?

Piping systems in many plants are generally more likely to be ignored than other equipment such as furnaces, reactors, heat exchangers, other pressure vessels and storage tanks. Piping between units within major chemical facilities can even have areas where the responsibility of ownership is clouded.

The Institution of Chemical Engineers understands the problems that can occur in poorly designed, poorly operated, and poorly maintained piping systems and offer an awareness training package entitled Module 12 'Safer Piping' which includes a group leader's guide, a video tape, and 105 colour 35 mm slides (Institution of Chemical Engineers, Hazard Workshop Module 012).

The 'Safer Piping' module states that piping systems typically represent 30% of total plant capital costs and that property insurance statistics claim that 40% of all major process plant losses are due to piping failures.

Piping systems may be relatively simple or complex and may convey utility services, or chemical fluids which can be highly flammable, highly toxic, corrosive, erosive or any combination of these properties. Piping systems within chemical plants handling hazardous fluids must be designed, fabricated, operated, and maintained considering the materials of construction, routing, support, thermal expansion, overpressure protection and vibration. Additional consideration should be given to small threaded piping, methods of pipe support, dead legs, overpressure due to trapped liquids, heat-up and expansion, external corrosion and corrosion under insulation.

But all of these obvious technical pipeline considerations are sometimes overlooked and many of the most infamous incidents have involved poor practices in piping design, in operations or in piping maintenance systems.

The American Insurance Association states:

Piping hazards are often underrated as compared to pressure vessels although the history of failures does not warrant this . . . Piping systems should be located in serviceable areas. They should be constructed in accordance with appropriate code requirements and inspected by appropriate means . . .' (American Insurance Association, 1979)

Specific piping system problems

The Institution of Chemical Engineers (1991) published 'A thirty year review of large property damage losses in the hydrocarbon–chemical industries'. This information was developed by Marsh & McLennan Inc., Protection Consultants.

Ten of the 19 major refinery incidents reported from 1981 to 1990 in the large property damage report were piping failures. Several recent examples of these piping failures which resulted in massive fires or explosions follow.

An 8 inch pipeline ruptures and an explosion occurs – 24 December 1989

The record sub-freezing weather may have contributed to the rupture of an 8 inch (20 cm) pipeline transferring ethane and propane 700 psig (4800 kPa gauge) in Baton Rouge, Louisiana, US, in December 1989. A vapour cloud was formed and ignited, and the powerful explosion broke windows 6 miles (10 km) away and was felt as far away as 15 miles (24 km). The blast destroyed a number of pipelines and the resulting fire involved two large diesel storage tanks holding about 3.6 million gallons (16 000 m^3) of diesel fuel. Twelve smaller tanks containing a total of 880 000 gallons (4000 m^3) of lube oil were also involved in the fire (Institution of Chemical Engineers, 1991).

The initial explosion created a partial loss of electricity and steam, and damaged the 12 inch (30.5 cm) fire-water pipeline connected to the dock fire-water pumps. The remaining fire-water pumps and municipal fire trucks drafted from other sources including the river. Over 200 fire brigade members and 13 pumpers were employed during the fire-fighting effort. About 48 000 gallons (220 m^3) of AFFF foam-type concentrate was used during the 14 hours of fire-fighting. Property damage estimates were $43 million (US, 1989).

A 2 inch high-pressure hydrogen line fails and the fire topples a reactor – 10 April 1989

In Richmond, California, a 2 inch (5 cm) line carrying gaseous hydrogen at a pressure of 2800 psig (19 300 kPa gauge) failed at a weld. The jet of fire impinged on the supports of a 100 ft (30 m) tall reactor in a hydrocarbon unit and the heat weakened the steel skirt on the large reactor.

The skirt, which was about 10 ft (3 m) in diameter, collapsed and the reactor fell, damaging associated equipment, including fin-fan coolers. Property damages were estimated at about $90 million (US, 1989), and it was estimated to take about 2 years to repair (American Insurance Association, 1979).

An 8 inch elbow ruptures from internal corrosion and a blast results in worldwide feedstocks disruptions

A major refinery in Norco, Louisiana, US experienced an explosion on 5 May 1988, resulting from a rupture of an elbow in process piping. Property damages were estimated to be $300 million (US, 1988), which was the most costly refinery incident up to that time. (The Phillips incident in Pasadena, Texas just 17 months later in October 1989, experienced property damages 2½ times greater.) This Norco incident had enormous effects on worldwide feedstock supplies (Institution of Chemical Engineers, 1991).

Operations appeared normal at this 90 000 barrel per day refinery until an 8 inch (20 cm) carbon steel elbow failed due to internal corrosion. An estimated 10 tons (9000 kg) of C_3 hydrocarbons escaped during the first 30 seconds between failure and ignition. The failed elbow was 50 ft (15 m) in the air and the large vapour cloud was believed to have been ignited by a nearby heater.

The blast destroyed the cracking unit control room and the most severe damage was in the 300 ft (90 m) by 600 ft (180 m) unit. The overpressure from the blast may have been as high as 10 psi according to analysis of bolt stretching on the towers (Institution of Chemical Engineers, 1991).

Piping design failure incidents – lack of remote operated emergency valves

The Feyzin BLEVE of January 1966 was one of the worst petroleum accidents in France and poor piping design was a major factor. 'BLEVE' is an acronym for a boiling liquid expanding vapour explosion. This phenomenon occurs when a pressure vessel containing liquid above its normal boiling point bursts due to fire exposure. Under these conditions, the escaping flammables form a vapour cloud which instantly results in a fireball.

Over 15 men were killed (most were fire-fighters) and about 80 people were injured during this incident. Two 1200 m³ spheres, 48 ft (14.6 m) in diameter, containing pressurized propane stored at atmospheric temperature, BLEVEd. In addition, three other spheres and two other pressure vessels burst and other petrol and fuel oil tanks burned. The severity of a sphere explosion propelled pieces of steel weighing up to 100 tons about three quarters of a mile (1.2 km) away. Property damages were about $18 million (US, 1966) (Institution of Chemical Engineers, 1991; Kletz, 1988c).

An uncontrolled release of butane from a 2 inch (5 cm) water drain manifold initiated the tragic event after an operator opened a valve and a blockage cleared. Poor drain valve and sampling valve piping design was a major contributor to this accident. Recommendations included installing remotely operated emergency valves on the drain line, restricting the size of the drain valve to 3/4 inch, and supporting the line (Kletz, 1988c).

Six years later, in March 1972, in Rio De Janeiro, Brazil, a similar accident occurred. Water was being drained from a 10 000 gallon sphere through a 2 inch gate valve with piping pointed to the ground. The operator left with the water draining from a fully opened valve. When he returned, liquid LPG was jetting out from the 156 psig (1075 kPa gauge) pressure on the sphere and splashing on the ground; the operator was unable to reach the valve. A poor piping design which lacked remotely operated emergency isolation valves contributed to this tragedy.

An 8 inch line ruptures in Mexico City and over 500 people die

The worst Western Hemisphere petrochemical incident, in terms of human suffering, resulted from the rupture of an 8 inch (20 cm) LPG pipeline. This incident occurred in November 1984, within a LPG terminal. The facility was supplied by three refineries in a suburb of Mexico City. The LPG was stored in six spheres up to 630 gallons (2400 litres) each and 48 bullet tanks with a total inventory of about 3.8 million gallons ($14\,400\,m^3$) at the time of the explosion and fire. Property damages were estimated at $19.9 million (US, 1984).

The exact cause of piping failure was not identified; however, a gas cloud which covered a large area ignited resulting in a 1200 ft (360 m) fireball. This started a chain of problems which resulted in BLEVEs or ruptures of four spheres and 44 cylindrical tanks. Some tanks weighing up to 20 tons sky-rocketed and landed 3900 ft (1200 m) away. Reports indicated that 542 people were killed in this catastrophe, 4248 were injured, and about 10 000 were left homeless. Unofficial estimates of human suffering were higher (Institution of Chemical Engineers, 1991; Kletz, 1988c).

References

American Insurance Association (1979) *Hazards Survey of the Chemical and Allied Industries*, Technical Survey No. 3, Engineering and Safety Service, AIA, New York, NY, p. 65

Fire Protection Association (1975) Anatomy of a disaster. Fire Prevention No. 110. *Journal of the Fire Protection Association* (August), pp. 12–20

Griffin, M.L. (1989) Maintaining plant safety through effective process change control. 23rd Annual Loss Prevention Symposium, American Institute of Chemical Engineers, Houston

Institution of Chemical Engineers. *Preventing Emergencies in the Process Industries*, Vol. 1, Hazard Workshop Module 006. Available as a training kit with a video tape, sixty 35 mm slides, a teaching guide, etc. from the Institution of Chemical Engineers, 165–171 Railway Terrace, Rugby CV21 3HQ, Warwickshire, UK, pp. 4.37–4.42

Institution of Chemical Engineers. *Safer Piping*, Hazard Workshop Module 012. Available as a training kit with a video tape, 105 35 mm slides, a teaching guide, etc. from the Institution of Chemical Engineers, 165–171 Railway Terrace, Rugby CV21 3HQ, Warwickshire, UK

Institution of Chemical Engineers (1991) *Loss Prevention Bulletin, Articles and Case Histories from Process Industries throughout the World,* **99**

Kletz, T.A. (1984) *Myths of the Chemical Industry, or 44 Things a Chemical Engineer Ought NOT to Know*, Institution of Chemical Engineers, Rugby, UK, pp. 27–28

Kletz, T.A. (1988a) *What Went Wrong? Case Histories of Process Plant Disasters*, Gulf Publishing, Houston, pp. 36–37

Kletz, T.A. (1988b) *What Went Wrong? Case Histories of Process Plant Disasters*, Gulf Publishing, Houston, p. 24

Kletz, T.A. (1988c) *What Went Wrong? Case Histories of Process Plant Disasters*, Gulf Publishing, Houston, pp. 113–117

Kletz, T.A. (1990) *Critical Aspects of Safety and Loss Prevention*, Butterworths, London, pp. 160–161

Lorenzo, D.K. (1990) *A Manager's Guide to Reducing Human Errors – Improving Human Performance in the Chemical Industry*, Chemical Manufacturers' Association, Washington D.C., p. 17

Occupational Safety and Health Administration (1990) *Phillips 66 Company Houston Chemical Complex Explosion and Fire – Implications for Safety and Health in the Petrochemical Industry*, OSHA, US Department of Labor, Washington D.C.

Warner, Sir Frederick (1975) The Flixborough Disaster. *Chemical Engineering Progress*, **71**, no. 9, pp. 77–84

Chapter 5

The one-minute modifier – small quick changes in a plant can create bad memories

Chemical plant managers aspire to hire individuals who can think for themselves and can analyse a complex problem and solve it quickly. However, it is risky business if these companies fail to properly educate all of their team members on the danger of changes. It is dangerous if companies allow well-meaning employees to introduce quick alterations into their carefully engineered plant without some sort of review. One-minute modifications included in this chapter can also include simple errors introduced by lack of understanding of the procedures, insufficient training, or failure to consult specifications. These small, easily implemented changes can create nightmares as exhibited in the following examples.

Explosion occurs after an analyser is 'repaired'

Several decades ago, an instrument mechanic working for a large chemical complex was assigned to repair an analyser within a nitric acid plant. He had experience in other parts of the complex, but did not regularly work in the acid plant. As part of the job, the mechanic changed the fluid in a cylindrical glass tube called a 'bubbler'. This 'bubbler' scrubbed certain entrained foreign materials and also served as a crude flow meter as the nitrous acid and nitric acid gases flowed through this conditioning fluid and into the analyser.

The instrument mechanic replaced the fluid in the bubbler with glycerin. Unfortunately, the glycerin was nitrated into nitroglycerin and detonated in less than two days. This dangerous accident resulted from an undetected 'one-minute' process change of less than a litre of fluid. It appears that a lack of proper training led to this accident.

Just a little of the wrong lubricant

While attempting to free a 2 inch (5 cm) plug valve in chlorine service which was becoming difficult to use, a chemical process operator located a lubricant gun that was marked 'for chlorine service' and contained a cartridge of the proper grease. Before the operator could disconnect the grease gun after he lubricated the valve, a mild explosion occurred. Evidence showed that someone had contaminated this gun with hydrocarbon grease. The rapid chlorination of the hydrocarbon grease created destructive pressures high enough to blow a ¾ inch (2 cm) hole in the bottom of the plug cock body and allow chlorine to escape from the system.

The chemical process operator did everything correctly. However, someone else in the organization apparently did not understand the dangers when he previously used the grease gun in hydrocarbon grease service.

Instrument air back-up is disconnected

A maintenance supervisor decided to temporarily use instrument air as a source of breathing air. For the safety of the workers, he disconnected the automatic nitrogen back-up system as a well-intended 'one-minute modification'.

The nitrogen back-up system was never reconnected, so later, when the instrument air compressor tripped, there was no auxiliary pneumatic power available. Without the instrument air back-up system capabilities, the process crashed down, and several batches were ruined (Lorenzo, 1990).

An operator modifies the instrumentation to handle an aggravating alarm

A chemical process operator taped down the acknowledge button on the instrument cabinet to stop an incessant alarm after a level alarm switch had slipped. The alarm switch slipped down its support into a sump and alarmed every few minutes as the sump pump cycled on and off. The acknowledge button, common to a number of alarms, deactivated the whole system (Lorenzo, 1990).

Because of this modification, the critical high-temperature alarm was not able to be sounded. The operator failed to see the alarm light and the situation eventually led to a large toxic vapour release (Lorenzo, 1990).

A furnace temperature alarm is altered

A major fire erupted in a non-flammable solvents plant after a 6 inch (150 mm) diameter furnace tube ruptured from overheating. At least 1800 gallons (6800 litres) of a combustible heat transfer fluid spilled and burned intensely, damaging four levels of structure within about 25 minutes. The emergency squad quickly responded, but this short-lived incident ended up costing over $1.5 million in direct property damages and over $4 million in business interruption (costs in 1979 US dollars) (Sanders, 1983) (Figure 5.1).

Figure 5.1 Firefighters cool down area in solvents unit

Solvents were being manufactured in five similar 9 ft (2.7 m) diameter gas phase reactors. This reaction is exothermic and requires a combustible heat transfer fluid for cooling. But, a single gas-fired, start-up furnace is shared between all five reactors to heat each reactor to start-up conditions.

Investigators determined that a chemical process operator made an error during a hectic day of operations. The operator tried to start up a natural gas-fired oil furnace without having established heat transfer oil circulation through the tubes. When the damaged equipment was examined, it was discovered that a rather primitive alarm system was still functional, but that someone had inappropriately increased the alarm set point to a temperature above the failure temperature of the tube. Perhaps the instrument system had given troubles, been a nuisance and had become a victim of a one-minute modification (Sanders, 1983).

Afterthoughts

Shortly after this incident a major campaign was initiated to improve the reliability of critical instrumentation systems. A 'Proof-test' programme was conceived. It took over seven years to identify all critical instrument systems in a 250 acre plant site, to improve or develop instrument loop sheets, to evaluate critical instrument set points, to calculate an acceptable test frequency rate and to get all 1800 systems properly tested.

The wrong gasket material creates imitation icicles in the summer

Molten caustic soda at about 700°F (370°C) sprayed and dripped for about 2½ hours, melting insulation on wiring and destroying plastic electrical

Figure 5.2 Molten caustic soda leaks because of an improper gasket

Figure 5.3 Caustic soda icicles are seen in June as 15 tons of material leaks

junction boxes and instrument control boxes (Figure 5.2). An estimated 15 tons (13 600 kg) of material slowly leaked out of the concentrator. The 98% caustic soda formed icicles hanging from the structure and molten piles of material accumulated on the ground (Figure 5.3). About $40 000 (US, 1988) was spent to clean up and recover from the damages. There were no injuries; there were no environmental insults.

This situation was inadvertently created when someone installed a compressed asbestos gasket, the nearly universal gasket for this chemical complex, on a thermocouple flange in the vapour space. Specifications dictated a metallic nickel gasket but a compressed asbestos gasket was installed in error. The improper gasket may have been installed months or perhaps years before, and it was usually exposed to a high vacuum and caustic soda vapours. Problems within the unit created a high liquid level which exposed the gasket to atmospheric pressures and molten liquid 98% caustic soda.

Another gasket error

A serious oil refinery fire resulted from the installation of the wrong gasket on a heat exchanger. The unit was operating normally, processing a very flammable naphtha about six weeks following an outage in which the exchangers were cleaned and tested (American Petroleum Institute, 1979).

A spiral-wound metal asbestos gasket was required by specifications, but a compressed asbestos gasket was installed. The gasket failed, causing a leak followed by a fire. The heat from the initial fire caused an adjacent hydrogen exchanger flange to open. Vapours flashed and impinged on an 18 inch (46 cm) hydrogen–oil line which then ruptured. This 18 inch (46 cm) piping failure released oil and hydrogen under 185 psig (1275 kPa) pressure (American Petroleum Institute, 1979).

The fire was brought under control after about 2 hours, but it was allowed to burn another hour and 20 minutes to consume the remaining fuel. The fire damages affected the electrical and instrument systems, piping, and structural steel equipment.

As a result of this fire a number of measures were taken to prevent a recurrence. Preventative measures included a review of gasket installation and purchase specification procedures. Drawings appropriate for turnaround work were developed to conform to new facility practices and have flange ratings marked. These measures were reviewed with turnaround planners and supervisors. A retraining programme covering gasket installation procedures for craftsman was conducted. Purchasing specifications for gaskets now require colour coding (American Petroleum Institute, 1979).

As compressed asbestos gaskets are phased out other leaks will occur

Compressed asbestos gaskets have been available for over 120 years. They are resistant to high temperatures. They are inert to many alkalis, acids, oxygen, high-temperature steam, petroleum products and organic solvents. In general, they are strong and relatively inexpensive. In many plants they have been the universal type of gasket.

Asbestos fibres are tightly bound in such gaskets. But if these gaskets are improperly handled, as, for example, by tearing them up with a power wire brushing or sandblasting them from adhering flanges, asbestos fibres can become airborne. Airborne asbestos fibres and dust are a known carcinogen linked to lung cancer, asbestosis, pleural plaques and other serious threats to the health of the employees handling the asbestos materials. Governmental agencies in many industrial countries have developed legislation to phase out the manufacture and distribution of compressed asbestos gaskets.

It is disappointing, but there is no universal substitute for the compressed asbestos gasket. In many cases, it will require two, three or more different, limited-application substitutes. Many substitute gaskets will cost between three and ten times more than the compressed asbestos they replace.

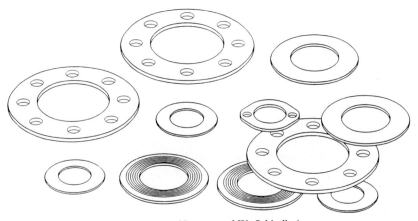

Figure 5.4 An assortment of gaskets (Courtesy of W. Schindler)

In some plants there will be a need for relatively expensive flexible graphite-type gaskets for high-temperature services. A Teflon-like gasket, which is also expensive, will be required for a wide variety of services such as petroleum-based solvents, oxygen and chlorine. A less expensive synthetic fibre with a rubber binder is available for less demanding services. Unfortunately, if certain gaskets are accidentally interchanged, chemicals could unexpectedly leak. A serious need is arising to spend more time to educate all maintenance personnel on details of gasket selection and installation as asbestos materials are phased out.

Other piping gasket substitution problems

A petroleum refinery was in the process of phasing out compressed asbestos gaskets to conform to new US laws. Asbestos gaskets had been used for years without any significant problems. During the conversion the company mistakenly assumed that the non-asbestos substitute gasket would seal the same fluids as the asbestos products if the gaskets were used within the manufacturer's temperature limitations.

Regrettably a refinery worker was injured by sulphuric acid because a 4 inch (10 cm) piping gasket failed. The acrylic fibres in the non-asbestos gasket material were attacked by the acid. The asbestos fibres had resisted the chemical attack.

In a different incident, within a chemical plant, the introduction of the wrong gasket in piping flanges extended a planned outage by an additional week due to an unplanned fire. A reactor was taken out of service for an outage and before start-up it was necessary to replace some 6 inch (15 cm) piping gaskets on the vessel.

The chemical plant stocked both a stainless steel spiral-wound gasket with a flexible graphite filler and a similar spiral-wound gasket with a mica–graphite filler. The gaskets looked similar, but did have distinctive colour coding identification on the guide rings. The out of spec gasket was installed.

As the reactor was being started up, flammable fluids leaked. The flammables ignited. The resulting damage was severe enough to delay the start-up by a week.

New stud bolts fail unexpectedly

About 20 years ago, a thrifty purchasing agent for a large chemical complex was concerned about the high prices of stud bolts that were coated with a plastic-like finish to inhibit rusting. He learned that cadmium-coated studs and nuts were available at more economical prices.

Some cadmium-plated bolts and nuts were purchased and installed as a test in the plant. More bolts were purchased and eventually, some of the cadmium-plated bolts were installed on the piping of a cracking furnace. Some high-temperature outlet flanges on the furnace started to weep a bit. A mechanic attempted to snug up a few bolts and they snapped due to a corrosion phenomenon called 'liquid metal embrittlement'. The alert mechanic informed his supervisor of the failures. The furnace was shut down and bolts were replaced with the properly specified high-strength stud bolts prior to any serious releases.

If there are several grades of bolts, gaskets, tubing, and fittings available, and if interchanging these fairly low-cost items can result in leakage or loss of containment, then only the material which is suitable for all services should be stocked in the plant. Only one short-duration fire or hazardous material release in a chemical plant could easily wipe out many years of savings realized by alternative lesser quality bolts, gaskets and fittings.

Hurricane procedures are improperly applied to a tank conservation vent lid

Hurricane winds and the accompanying intense long-lasting showers frequently threaten US Gulf Coast chemical plants. It is not unusual for an area to prepare for one or two of these potentially dangerous storm systems during certain summers.

There were perceptions at this plant site that hurricane-force winds would open conservation vacuum/pressure lids on large atmospheric tanks and torrential rains would enter the storage tanks, contaminating a moisture-free product. To protect certain low-moisture specification chemical products, a chemical plant tethered (or tied down) the hinged vent lids for limited opening to allow movement while reducing the chances of intrusion of rain. Apparently, the plant failed to train the individual, who fully restrained a 24 inch (61 cm) lid before Hurricane Juan was scheduled to visit in 1985 (Sanders *et al.*, 1990).(See Figures 5.5 and 5.6.)

It was really unnecessary to tie down this conservation lid, as this 100 ft (30 m) diameter tank was out of service, empty and blinded on all process lines. A nitrogen inerting system was left in service, and the tank was left in that condition for several weeks.

A cold front passed through the area at the end of September. The nighttime temperatures dropped to about 50°F (10°C) and the daytime temperatures rose to the high 80s (about 30°C). Assume that the nitrogen inerting system pressure was set for 0.5 oz/in^2 gauge (22 Pa). If the gases in the tank cooled down to about 55°F (13°C) at the coolest part of the night, and the mid-day temperature within the tank reached 85°F (29.5°C), then pressures of over 14 oz/in^2 gauge (600 Pa) could be expected in a leak proof tank system. This 14 oz/in^2 gauge pressure would have been nearly eight times the mechanical design pressure of the 100 ft diameter tank (Sanders *et al.*, 1990).

The mid-day warming weather conditions created damaging overpressure within the tank, which required significant repairs to the cone roof and the adjacent side wall. Significant repairs were required, including the replacement of a portion of the upper vertical wall and rewelding most of the seams on the roof. The necessary repairs to this tank cost about $56 000 (US, 1985) (Sanders *et al.*, 1990). (See Figure 5.7).

Afterthoughts on damages to the tank

In specific extraordinary situations it may be necessary to temporarily increase the set point, or bypass or deactivate some type of instrumentation or other equipment safeguard. However, in the 1990s this should only be done in an established manner, according to appropriate procedures with proper approval. Furthermore, any special defeat to a safeguard should be properly signalled for the operators and recorded in a special log that is regularly reviewed, so that the defeated system can be properly restored at the earliest possible time.

Painters create troubles

Sandblasters and painters have occasionally produced troubles for chemical processing with one-minute modifications by getting blasting sand

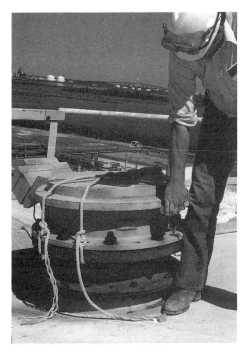

Figure 5.5 Hurricane precautions are improperly taken to prevent water intrusion

Figure 5.6 Overpressure problems created by a one-minute modification and changing weather

Figure 5.7 Repairs in progress on a large tank

or paint in moving parts of pump seals, safety relief valves, etc. There is always the predictable painting over of the sight glass windows or identifying tags.

However, overzealous painters can also create other problems when they paint over the fusible links of sprinkler heads or fire protective system pilot heads. An innocent-looking coat of paint quickly elevates the sprinkler head's melting temperature, and allows a fire's destructive temperatures to rise above the specified detection level. Paint can hinder the release of the fire protection system signal air or hinder release of fire-water spray.

Pipefitters can create troubles when reinstalling relief valves

For totally unexplainable reasons, a mechanic reinstalled a small threaded safety relief valve (SRV), using pipe fittings which reduced the discharge from 2 inch piping to 1½ inch piping (Figure 5.8). If this SRV were to relieve the compressed air it was designed to relieve, then the back pressure would cause it to rapidly slam shut. It would pop open again and quickly slam shut again. Within a very short time the valve might self destruct.

There have been reports of pipefitters and other craftsmen assembling the discharge piping into the outlet side of a threaded safety relief valve improperly. In such cases, if the threaded SRV opened, the thrust could unthread the SRV out of the connecting nozzle.

Figure 5.8 A threaded safety relief valve reinstalled with a reduced discharge piping arrangement

Another pipefitter's error

Pipefitters have occasionally installed reverse buckling rupture discs upside down. Depending on the rupture disc manufacturer specifications, this usually means that the bursting pressure of the disc will be 50% higher than the value the manufacturer guaranteed as the relieving pressure. Such incidents must have happened at numerous plants and have been whispered about, but seldom documented.

Rupture discs are often used as the primary or secondary overpressure devices on reactors, distillation columns, tanks and piping. In many chemical manufacturing companies, rupture discs are installed upstream of safety relief valves, especially if the system contains corrosives, flammables, carcinogens, or toxic materials. The earlier designs of rupture discs allowed operation of the system up to 70–80% of the rupture disc rating by using a disc that was pre-bulged away from the pressure source.

Newer rupture disc designs, and especially 'reverse buckling types,' allow operation of pressures up to 90% of the rupture disc ratings, resulting in fewer nuisance premature failures and associated leakage. Reverse buckling discs are designed to be installed with the bulged surface projecting into the pressurized fluid. They are widely used, but occasionally they are mistakenly installed upside down, which significantly increases the relieving pressure. To ensure the proper installation of rupture discs, the identification tag must be visible when looking toward the pressurized system.

The simple error of installing a rupture disc upside down is easy to do in a minute or two and can jeopardize the integrity of the overpressure protection system.

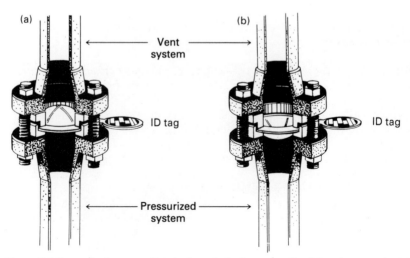

Figure 5.9 Pressurized system. Standard pre-bulged rupture disc (a) and reverse buckling rupture disc (b) installation (Courtesy of Oklahoma Safety Equipment Company)

A cooling water system is safeguarded and an explosion occurs some months later

David Mansfield, in an article in the British *Loss Prevention Bulletin*, captures the focus of this book in his introductory paragraph: 'Every once in a while an apparently subtle change to a process or method of operation gives rise to a new or at least totally unexpected hazard' (Mansfield, 1990).

The management of a large research establishment was concerned that it should treat its cooling water system, which was a part of its air conditioning system, to prevent the dangers of the widely-publicized incidents involving the legionella bacterium, or Legionnaire's disease. Employees dosed the water in the system with a suitable biocide (Mansfield, 1990).

Some months after installing a biocide dosing system on the cooling water unit, one of the dosing tanks overpressurized and ruptured. The tank contents were propelled about 20 ft (6 m) in the air. The unusually high temperature of the tank was a clue that some form of exothermic chemical reaction had occurred. Liquids and gases in the vicinity were sampled and analysed, and this revealed the presence of chlorine, bromine and their acid gases, which were probably released during the explosion.

After the accident, the cooling water was found to contain 14% ethylene glycol (antifreeze). Tests were conducted with the biocide and the ethylene glycol solution. It was determined that after a 'dormant period' of several days, the temperature of the mixture suddenly increased and the solution released bromine gas. It was also determined that at elevated temperatures of about 60°C (140°F) the reaction occurred without the delay.

The ruptured tank had been connected to the cooling tower system that had not been operated since the start-up of the biocide system. This stagnated condition and the unusually high ambient temperatures contributed to the explosion (Mansfield, 1990).

Lack of respect for an open vent as a vacuum relieving device results in a partial tank collapse

Two fairly large, dished-head, low-pressure 32% hydrochloric acid tanks supplied a periodic flow of acid to a chemical process (Figure 5.10). The two rubber-lined, horizontal storage tanks were 12 ft (3.7 m) in diameter and about 40 ft (12.2 m) in length. Both vessels were vented to a small scrubber which used water absorption for any fumes released during the filling operation.

The scrubber was a small vertical vessel about 28 inches (71 cm) in diameter and over 12 ft (3.7 m) high and contained ceramic packing. This flat-top scrubber had a 4 inch (10 cm) hole which served as a vent. This vent allowed air to escape the system during the filling operation and allowed air to enter to displace the volume of acid periodically pumped to the process with a 50 gallon min^{-1} (11.4 m^3 h^{-1}) pump.

The small scrubber was believed to be presenting some operational difficulties and may have been partially plugged. A supervisor climbed a ladder, looked into the vent opening and observed an extensive growth of

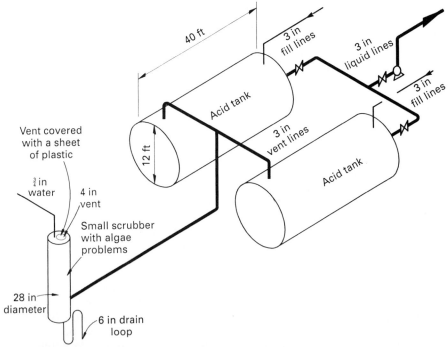

Figure 5.10 Acid system before tank is partially sucked in (Courtesy of J. M. Jarnagin)

algae. The supervisor reasoned that the algae must be part of the operational problems and if he blocked off sunlight the algae would die. He merely placed a small opaque PVC plate over the opening to stop sunlight from entering the scrubber. He unknowingly modified the vent; air could not flow into the tanks as they were pumped out and the top of one tank sucked in due to a partial vacuum.

Lack of respect for an open vent as a pressure relief device costs two lives

A horizontal tank was mounted above a building which housed a blending and packaging operation. The tank was used to store a wax-like material which had a melting point of about 212°F (100°C). Each evening the tank was recharged for the next day's operation.

This cylindrical, low-pressure, horizontal tank was designed for about 5 psig (0.3 bar) and was kept hot by a steam coil. The transfer line to the tank was steam traced and insulated to keep it clear, but the heat tracing and insulation system on the inlet transfer piping was less than ideal. The inlet piping had to be dismantled on several occasions to clear it because of solidificaton in the line (Figure 5.11) (Kletz, 1988a).

The procedure was to clear the transfer line of the waxy material with a compressed air supply just before the transfer line was put in service. On

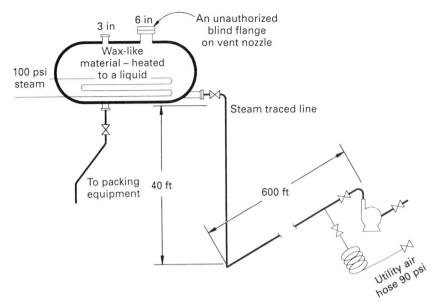

Figure 5.11 Storage tank explodes before a transfer (Courtesy of J. M. Jarnagi.ı)

the evening of the accident, an operator connected an air hose to a 90 psig (6 bar) supply to clear the entire 200 m (600 ft) of feed line. The operator then walked to the elevated tank to determine if air could be heard rushing out of the tank vent. There were no sounds heard from the tank vent. The operator spoke to two maintenance employees working near the end of the 20 m³ (700 ft³ or 5300 gallon) tank and left the area to find his supervisor. Shortly after that, one end of the tank blew off and caused the two maintenance mechanics to plunge some 40 ft to their death (Kletz, 1988a).

According to the engineering drawings, a 6 inch (150 mm) diameter open nozzle was originally considered to be the tank vent. Over the seven years of operation, this 6 inch flange had been blinded and an unused 3 inch (75 mm) nozzle served as the vent. No one knew when this 'one-minute modification' had occurred. Experience elsewhere in this plant with many similar products had established that a 3 inch vent hole could block solid with this wax, but it was most unlikely that a 6 inch nozzle would clog (Figure 5.11) (Kletz, 1988a).

Afterthoughts

An open nozzle can be the primary relief device and should be treated with the same respect as a safety relief valve. However, this plant, as well as many others, is not treating open vents with the same respect as safety relief valves. No process foreman or chemical process operator would dream of asking a fitter to replace a relief device with a smaller one without requesting that calculations be performed (Kletz, 1988a).

An open vent which serves as a relief device should be registered for regular inspection to ensure that it is not obstructed. It is also obvious that any open vent designed for overpressure protection should not be designed so that it can be easily blinded (Kletz, 1988a).

The misuse of hoses can quickly create problems

Hoses are necessary to purge vessels, to wash vessels, to compensate for unavailable equipment, to help make transfers of fluids, to prepare for maintenance, to vent systems, etc. When hoses are misused, many unpleasant things can occur, including some of the world's worst chemical plant accidents.

All plants have various utility fluids, such as well water, compressed air, low-pressure steam, and nitrogen, and anyone who has opened a valve on a utility service expects the utility fluid to flow from the utility header and out of the hose. However, if a utility hose is connected to a higher pressure system and there are not any check valves, undesirable chemicals can back into the utility system and exit in unexpected places.

Some of the many unpublished errors created with hoses

Many stories have been repeated in control rooms and at inter-plant meetings where flammables have backed into the nitrogen headers, or where caustic soda, salt water, cooling tower water, etc., has backed up into the drinking water in the 1960s and 1970s. One such 1960s incident involved a sandblaster using a hood receiving breathing air from a plant compressed air network. This painter suddenly received a slug of ammonia-tainted air in his hood while working early one morning. The night before this incident, the compressed air was being used to pressurize a large condenser which was about 4 ft (1.2 m) in diameter and 30 ft (9 m) long.

According to the instructions to the operator, he should pressure up the ammonia condenser to compressed air system pressure, block off the air supply and discharge ammonia fumes to the atmosphere. The condenser was pressurized to full air system pressure, but unfortunately it was not disconnected. Therefore, at the beginning of the day, when the arriving maintenance mechanics and operators put a high demand on the air system, the condenser acted as a surge tank. As the system pressure dropped, undesirable ammonia fumes travelled to the sandblaster's hood.

Another undocumented story concerns a dangerous quantity of hydrazine (a very toxic vapour, corrosive to the skin and spontaneously igniting flammable liquid) backing into a power plant's drinking water supply from a utility well water hose connected to the boiler water treatment system. This reportedly resulted in a number of power plant workers being hospitalized.

In another tragic incident, a piping contractor was working within a chemical plant which was completing a major modification. The role of this piping contractor was the fabrication and installation of long runs of 24 inch (61 cm) diameter glass fibre headers. The contractor had nearly completed the job and crawled through the piping for an inspection.

During this inspection it was recognized that additional glass fibre work was required. These repairs would require grinding to properly finish one of the field made joints within this header. The contractor generally used a pneumatic grinder for these types of repair and this chemical plant was equipped with many compressed air utility stations which could be connected to nominal ¾ inch (1.9 cm) hoses.

The plant was also engineered with compressed nitrogen utility stations and to avoid accidents the plant equipped the nitrogen stations with different, non-interchangeable fittings. It seems that the contractor failed to understand why a utility station had different fittings, and modified the 'air' station to connect to a hose and a grinder.

Tragically, the repair job was many feet down the header and the nitrogen powered grinder exhausted into the confined space and asphyxiated the repairman. A well-intentioned simple piping one-minute modification by a small contractor cost the life of a family member co-worker.

Some of the most tragic and well-remembered accidents also had their origins in a mini modification made with a hose connection. Certainly, the Bhopal tragedy, the Three-Mile Island incident and the Flixborough disaster were each initiated by the improper use of hoses.

The water hose at the Flixborough disaster

The explosion in the Nypro factory at Flixborough, UK had a major impact on the entire chemical industry, and especially the European chemical industry. In short, this 1 June 1974 explosion was of warlike dimensions. The explosive energy was equivalent to the force of 15 tons of TNT. Twenty-eight men were killed at the plant and 36 other individuals suffered injuries. Outside the plant, 1821 houses and 167 shops and factories suffered some degree of damage (Warner, 1975).

Almost every discussion on the hazards of plant modifications uses this accident as an example of the need to control plant modifications. The authors, almost without fail, centre their attention on the 20 inch (500 mm) diameter piping that was improperly installed to bypass the no. 5 reactor, which had developed a 6-foot-long crack. But the reason for the crack is usually just briefly mentioned.

In fact, a simple water hose or the temporary routing of cooling water to the stirrer gland of no. 5 reactor was the precursory event of this tragedy. A water hose stream was directed to the gland to cool and quench a small cyclohexane leak. The cooling water contained nitrates. Unfortunately nitrates encouraged stress corrosion cracking, which created a crack on reactor no. 5 so that it had to be repaired (Warner, 1975). Pouring water over equipment was once a common way of providing extra cooling, quenching or absorbing leaking material.

Hoses used to warm equipment

Pouring of water over certain equipment to warm it has also been successfully employed in many chemical plants. In a plant that required a small amount of chlorine gas from a ton cylinder, operators have poured warm water over the vessel to slightly increase the chlorine vapourization rate. This has worked well for years.

However, there have been stories at inter-plant safety meetings that some chemical plants received more chlorine than they bargained for because hoses created dangerous modifications. It seems that individuals who were unfamiliar with one-ton chlorine cylinder design placed steam hoses exhausting directly onto the portable cylinders to increase chlorine vaporization.

Chlorine cylinders are equipped with three fuse plugs (which melt at 165°F or 74°C) in each end. The steam reportedly melted the plugs and the chlorine escaped, through ⅜ inch (0.9 cm) diameter orifices.

Three-Mile Island incident involves a hose

At the Three-Mile Island nuclear power plant in Pennsylvania, USA, on 28 March 1979, a hose contributed to the accident. In short, a nuclear reactor overheated and a small amount of radioactivity escaped and shattered the public confidence in the safety of nuclear power. It is believed by the technical community that no one is likely to be harmed by this release, but it led to retarded growth of nuclear power in the United States (Kletz, 1988b).

Full details of this accident can be found in several major reports. But, for purposes of considering just the 'one-minute modification' aspects of the accident, the trouble started when one of the parallel paths that treat the secondary water started plugging. When this resin polisher system, which is designed to remove trace impurities, started to plug, an operator decided to clear the blockage with instrument air (Kletz, 1988b).

Clearing any system with instrument air as a pressure source is a bad idea. Other pneumatic sources such as utility air, plant air or nitrogen should be used instead. However, the Three Mile Island instrument air was at a lower pressure than the water stream on the resin polisher system and, despite the presence of check valves, water entered the instrument air system, causing several instruments to fail, and the turbine tripped. Through a series of other errors the water covering of the radioactive core was uncovered, allowing an escape of a small amount of radioactivity. The result of this was a great public reaction and the US nuclear industry received a setback. If it were not for the improper hose connection, or the erroneous 'one-minute modification', it is unlikely that anyone outside of the Pennsylvania area would have heard of the Three Mile Island power station (Kletz, 1988b).

The Bhopal tragedy is initiated by use of a hose

The front page and feature article of the *Wall Street Journal* in July 1988 explains the Union Carbide belief that a disenchanted worker initiated a one-minute plant modification with a hose. The article is introduced with these three paragraphs:

'On a chilly winter night, a disgruntled worker at Union Carbide India Ltd. sneaks into a deserted area of its Bhopal pesticide plant where storage tanks hold thousands of gallons of a toxic chemical.

He removes a pressure gauge on one tank, attaches a water hose to the opening and turns on the faucet. Two hours after he steals away, a calamity he never foresaw begins unfolding. The tank rumbles. Vapours pass into a vent tower. A large cloud of poison gas drifts out into the night.

Such a chain of events, Union Carbide Corp. believes, explains the gas leak that killed 2,500 people and injured thousands more in Bhopal on Dec. 2 and 3, 1984. The company insists that sabotage – not sloppy corporate practices in the Third World – caused what has been called the worst industrial disaster in history.'
(Hays and Koenig, 1988)

Afterthoughts on 'one-minute modifications'

The chemical industry has an excellent personal safety record because it approaches the majority of its activities with a great deal of discipline; but the general public has even higher expectations of this industry. Significant process changes are being properly reviewed at most chemical plant facilities, but the simple things, such as proper lubricants, proper gaskets, tampering with safety equipment, and control of hose connections, is a day-to-day consideration that may not have received as much attention as it should.

Each plant must persistently train and create awareness in their employees of potential dangers when faced with even small changes that can be 'easily' introduced into their plant. This message must be given not only to the engineers and supervisors, but also to the chemical process operators and mechanics.

References

The majority of the material in this chapter was presented in a technical paper at the 25th Annual AIChE Loss Prevention Symposium in Pittsburgh, PA, US in August 1991. The paper was entitled 'Mini modification miseries – small quick changes in a plant can create bad memories,' and presented by Roy E. Sanders.

About 75% of this chapter was published as an article entitled 'Small, quick changes can create bad memories' in *Chemical Engineering Progress*, May 1992. The publisher, the AIChE, has given permission to freely use this material in this chapter.

American Petroleum Institute (1979) *Safety Digest of Lessons Learned*, Section 2, API, Washington D.C., pp. 94, 95
Kletz, T.A. (1988a) *Learning from Accidents in Industry*, Butterworth Scientific, Guildford, UK, Chapter 7
Kletz, T.A. (1988b) *Learning from Accidents in Industry*, Butterworth Scientific, Guildford, UK, Chapter 11
Hays, L. and Koenig, R. (1988) Dissecting disaster. *The Wall Street Journal*, Southwestern Edition, Vol. LXXXII, No. 4, 7 July, Beaumont, TX, p. 1
Lorenzo, D.K. (1990) *A Manager's Guide to Reducing Human Errors – Improving Human Performance in the Chemical Industry*, Chemical Manufacturers' Association, Washington D.C., p. 16

Mansfield, D. (1990) An explosion in a cooling water system. *Loss Prevention Bulletin, Articles and Case Histories from Process Industries throughout the World,* **94**, 25

Sanders, R.E. (1983) Plant modifications: troubles and treatment. *Chemical Engineering Progress* (February), 76

Sanders, R.E., Haines, D.L. and Wood, J.H. (1990) Stop tank abuse. *Plant/Operations Progress* (January), 63

Warner, Sir Frederick (1975) The Flixborough Disaster. *Chemical Engineering Progress* (September), 77–84

Chapter 6

Failure to consult or understand specifications

The value of up-to-date, practical specifications is well known to technical employees within chemical plants. But at times these specifications are not properly maintained or persistently communicated to other members on the plant team, to equipment suppliers or to construction crews.

Many of the incidents covered in Chapters 2, 3 and 5 could also be repeated in this chapter. The incidents that follow occurred primarily because clear, concise specifications were either not readily available or because no one took the time to consult these vital design or operating criteria.

Failure to provide operating instructions costs $100 000 in property damage

Up-to-date standard operating procedures are now required by US law for all chemical companies which handle significant quantities of highly hazardous chemicals. However, in the 1970s, not all chemical plants had up-to-date procedures to cover every situation.

An accidental fire interrupted the start-up of a US Gulf Coast plant during a cold winter morning. The plant experienced $100 000 (US, 1978) of property damages. Less than 50 gallons (200 litres) of combustible heat transfer fluid burned to create such extensive damage.

Lack of operating instructions and lack of rigid operating discipline were major contributing elements in this accident. Quick, decisive, co-operative efforts between operations and the emergency squad limited damages. Thankfully, there were no injuries.

Plans were underway to return a reactor to service. This gaseous phase reactor requires a start-up temperature in excess of 500°F (260°C). The start-up temperatures are achieved by circulating a heat transfer fluid through a natural-gas-fired heater and through the reactor tubes.

The heat transfer piping system is not simple. There were three reactors and the heat transfer system is used for both heating and cooling the reactor. The circulating fluid heats the reactor to several hundred degrees Fahrenheit to start up the reaction. Once the reactor starts, the circulation removes the heat from this very exothermic reaction (Figure 6.1).

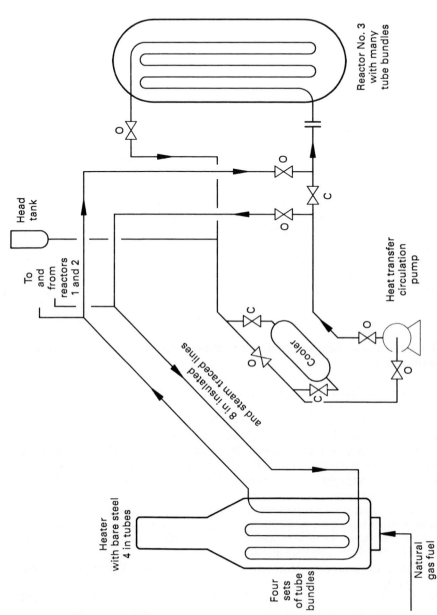

Figure 6.1 Heat transfer system just before the fire (Courtesy of J. M. Jarnagin)

The weather had been colder than usual at this southern US plant. The operating crew was attempting to establish flow through the loop from the discharge of circulating pumps, through the gas-fired heater, through the reactor and returning to the suction of the circulating pumps. Flow could not be easily established (Figure 6.1).

The entire 8 inch (20 cm) piping system was steam traced except for the heater tubes. It was assumed that the heat transfer fluid was frozen in the four heater tube passes. Each pass was a bare 4 inch (10 cm) diameter heater tube with five bends and the equivalent of 72 ft (22 m) of straight pipe.

The foreman and the operations team discussed the situation and decided to light a small fire in the heater to slowly thaw the material in the tubes. This method had been successful for a start-up several weeks before.

About 8:30 a.m. the burner was ignited. The fuel rate was adjusted to produce a flame between 1 and 2 ft (0.6 m) high. About 9:30 a.m., the operations foreman returned to this busy control room, and noticed that the natural gas flow to the heater had been significantly increased. Fuel was reduced to about one quarter of the existing flow.

Black smoke and fire from the heater stack was reported by a chemical process operator, while making his round at about 9:45 a.m. Nearby sprinkler systems were activated, and the emergency squad was summoned. The emergency squad set up fire hose streams and began cooling down the heater structure. Within 15 minutes all external flames from the heater ceased, but black smoke continued to pour from the heater stack. A few minutes later the fire flared up through the stack for about five minutes more.

The fire investigators could find no one, including the four operators, who admitted to increasing the natural gas flow between 8:30 a.m. and 9:30 a.m. Regrettably, the flow recorder on the natural gas supply to the heater was not properly inking. Therefore, none of the flow rates or times of gas flow rate change could be established.

It was easy to spot two failed tubes during an inspection of the heater after the fire. One of the 4 inch (10 cm) diameter tubes was swollen like a bubble with a split that was about 2.5 inches (6.4 cm) wide and 4.5 inches (11.4 cm) long. The other tube failure was smaller. The failures were about 6 ft (1.9 m) from the bottom bend of the tube.

What did the investigators find?

The investigators concluded that the following items or conditions contributed to the cause of the fire:

1. There were no written instructions for thawing out frozen tubes in the heater.
2. The operations crew was not fully aware of the hazards of lighting the main burner before a flow through the heater was established.
2. The system was not engineered with the complement of safeguards recommended by insurance guides.
4. The heater tubes had a carbon build-up inside the tubes which restricted flow to some passes. There was some external thinning of tubes in the higher heat flux zones.

Other thoughts on furnaces

The Institution of Chemical Engineers markets '*Furnace Fires and Explosions*' (Institution of Chemical Engineers, Hazard Workshop Module 005). One of the advertising leaflets for this workshop makes the following generalization:

> Furnaces are comparatively simple items of a plant, and because they are unsophisticated they tend to be imperfectly understood by operators, and plant managers alike. Their tolerance to abuse islimited, and once abused their useful life can be drastically shortened. Worse still they may fail suddenly, since furnace tubes distort easily andthen fracture. Such failure is often severe, with a consequential fire and/or explosion.

Low-pressure tank fabrication specifications were not followed

Many vertical low-pressure storage tanks containing flammable or combustible liquids are designed with a weak weld seam on the wall-to-roof connection. This loss prevention feature allows the roof to separate and peel back if there is an internal fire, an internal explosion or just a pneumatic overpressure situation within the tank. If the roof relieves or is blown off, the liquid is still contained within the walls if these tanks are properly designed and fabricated. These conical roof tanks are often built to the American Petroleum Institute (API) Standard 650 'Welded Steel Storage Tanks'.

There is one documented report that a welder had good intentions when he decided that the standard of welding for the roof-to-wall seam was inadequate. This well-meaning welder added a full-strength weld to join the roof to the walls (Kletz, 1990a). There were no reported problems on this tank, but the safety feature was made ineffective.

Another case history (Institution of Chemical Engineers, 1981a) discusses a dramatic failure of the wall-to-floor seam of an oil tank. A vertical lubricating oil storage tank, 6.1 m (19 ft) in diameter and 7.4 m (24 ft) high, failed catastrophically when it was accidentally overpressured by compressed air injected into the tank. The compressed air supply was designed to provide mixing within the tank. There were no injuries and no fire.

After the tank was destroyed, investigators found problems with the pressure vacuum vent valve as well as concealed problems created by 'non-spec' roof construction methods. The construction method provided extra roof-retaining strength. Although the roof-to-wall weld was correctly made, this safety feature was made ineffective due to some unexplainable internal brackets which were not on the fabrication drawings. The wall-to-roof brackets strengthened the roof, prohibiting it from an easy separation. Instead of a weak roof tearing away, the bottom seam of the tank split and lubricating oil suddenly dumped all over the area. The bottom seam tear was so massive that the lubricating oil rapidly drained and created a damaging vacuum on the tank which caused a partial collapse of the side walls.

Explosion relief on low-pressure tanks

Explosion relief for vapour–air combustion within a closed-top low-pressure tank cannot be accomplished by customary tank conservation vents or other lightweight protection. A well-written brief description of safety features of low-pressure tanks can be found in the *Loss Prevention Bulletin* (Institution of Chemical Engineers, 1981b).

An excellent understanding of design considerations of the weak roof can be found under the heading '1.3.5 *Explosion Relieving*'. A few paragraphs are as follows:

'In the cases of most tanks, explosion venting is accomplished by lifting of part, or all, of the roof. For this to take place the roof must be attached to the top peripheral angle by the smallest feasible weld. API standards specify that the weld be no greater than 5 mm (3/16 in). Preferably this weld should be of the order of 3 mm or 4 mm.

A tank must have other qualities for successful explosion venting. The cone roof slope must not exceed 1 in 6 and should preferably be no more than 3/4 in 12. If the roof is supported by roof rafters the roof plates must not be attached to the rafters but must simply rest on the rafters. Roof lifting must not be hampered by railings or other structures above the roof. The roof should be able to lift, unhampered, to a height sufficient to give a peripheral open area equal to 1½ times the tank cross section. The roof should be as thin as possible, preferably not over 6 mm (¼ in). Both roof release pressure and roof weight per unit area should be as small as possible to minimise [*sic*] energy development before roof lift.

The tank must not lift up at the bottom under any conditions . . . Thus tiedown bolts may be required . . . the primary objective of a weak seam roof is to ensure that the contents remain in the tank . . .'

There are many reports and photos of low-pressure tanks that experienced overpressure conditions or explosions sufficient enough to tear open the roof and yet the walls and the bottom retained their integrity (Sanders *et al.*, 1990; Institution of Chemical Engineers, Hazard Workshop Modules 003 and 006). Most of the associated fires were confined to the inside of the tank. (See Figures 6.2 and 6.3.)

Piping specifications were not utilized

Piping system changes occur daily within major chemical plants. This section covers five different incidents involving piping changes in which piping specifications were not followed.

A small piping modification – for testing the line – created a failure point

A leak was observed on a 1 inch (2.54 cm) connection above a discharge header in a refinery. While carefully removing insulation around the leak to better observe the leak source, the mechanic aggravated the situation, and a threaded nipple and valve blew off. A stream of hot 670°F (354°C) oil slurry sprayed the area and produced a dense cloud of heavy oil mist. The

Figure 6.2 After the top opened, a fire burns and the liquid is still contained (Courtesy of the Institution of Chemical Engineers)

spray of oil was propelled about 90 ft (27 m) or higher into the air and the mist covered an extensive area of the refinery (American Petroleum Institute, 1979).

Immediately after the oil mist cloud formed, a shutdown of the unit was initiated, and the alert team started the fire-water monitors. The mist ignited about 10 minutes after the threaded connection blew open and the fire was extinguished about 90 minutes later. Luckily, only a portion of the total escaping oil burned.

During a previous modification, a contractor installed the coupling and nipple for pipeline testing purposes. No refinery personnel were aware of this threaded connection. Refinery piping specifications called for removal of any such nipple and backwelding a threaded plug before the line was insulated (American Petroleum Institute, 1979).

Figure 6.3 A fire in a giant floating roof crude oil tank (Courtesy of the Institution of Chemical Engineers)

The potential of additional hidden hazards was well understood. Employees were questioned if there were more of these off-spec surprises. A two-part programme was instituted by the company with the objective of eliminating similar troubles.

A list of likely pipeline hazards was assembled, and a technical staff member accompanied the foreman in an inspection tour of each operating area. The unit foreman assisted in the inspection. The list of 21 items for the self-inspection programme can be found in American Petroleum Institute (1979). This practical list included:

1. Inadequately supported small piping and pressure gauges
2. Vents, drains and other connections with no obvious function
3. Large threaded piping
4. Cast iron or brass fittings or valves in hydrocarbon service
5. Piping and equipment which were still connected to process and not in use
6. Improperly supported piping

The second part of this follow-up activity involved review and update of all of the engineering drawings to ensure that the miscellaneous vents, drains, sample points, etc. are documented. The fire investigation team also directed attention to the piping inspection programme to ensure that all lines that could cause an immediate fire in case of failure were regularly inspected. Chapter 8 covers inspection programmes.

Correcting piping expansion problems

A major food, detergent and consumer products company had defined some piping expansion problems. The solution seemed obvious and several flexible elastomer expansion joints were selected and installed to meet the process conditions within a soap manufacturing plant (Griffin, 1989).

One of the expansion joints accidentally failed while being cleared with a steam supply. Upon investigation, it was learned that the sponsors of this modification did not address the fact that sometimes process lines were blown clear with 150 psig (1040 kPa gauge) steam. The robust piping system was compromised, and could not withstand the heat. Fortunately, there were no injuries.

Failure to follow piping specifications as piping supports are altered

The environment was hostile to most improperly painted metals; corrosion threats were ever present in this chemical solvents reactor area. The reactor piping system was designed to be able to handle the temperature extremes between ambient to about 700°F (370°C), and the piping was engineered with spring supports and expansion joints to minimize stress on equipment during the heat up and cool down (Sanders, 1983).

However, in the late 1970s the operations and maintenance team of the solvents unit did not respect the need for these spring hangers. As the spring hangers became impaired, they were just removed, and the piping was solidly supported (Sanders, 1983).

This decision was a serious mistake. Many heat exchangers in this area were manufactured of brittle graphite. Other problems in January 1981 allowed a vaporized, combustible heat transfer oil into the system which normally only contained a non-flammable solvent and acidic gases. A graphite condenser fractured, and a vapour cloud of combustible heat transfer fluid was released and ignited. The resulting intense fire created property losses of over 500 000 dollars (US, 1981); however, there were no injuries. (See Figures 6.4, 6.5 and 6.6.)

The incident occurred late in the day shift and the incoming evening emergency squad assisted the day shift crew in supplying about 6000 gallons $\min^{-1}$ (380 litre s^{-1}) of fire hose coverage in addition to fixed fire water deluge systems. The fire was out in about 25 minutes.

The small amount of maintenance required to remove the troublesome spring hangers was not considered a change, because the team did not understand the specifications.

Piping systems substitutions damage a six-stage centrifugal compressor

A large six-stage dry chlorine centrifugal compressor was severely damaged during a start-up in 1990. Moisture entered this 9-foot-long piece of precision equipment and ferric chloride corrosion products quickly formed which plugged up this machine, halting the production of over 600 tons per day of product.

For a period of time just after start-up, the piping upstream of this compressor operated under a vacuum. This upstream piping was equipped with a set of double block and bleed valves to a scrubber which circulated diluted caustic soda. During this unusual start up configuration moisture was sucked into the system and the wet chlorine completely dissolved the stainless steel ball within 3 inch (7.6 cm), quarter-turn valves.

Figure 6.4 Fire in solvents plant as seen from the road (Courtesy of G. O. Heintzen)

Figure 6.5 Fire intensely burning in the reactor area in the third level of the structure (Courtesy of G. O. Heintzen)

Figure 6.6 The emergency squad responding to a fire in the structure (Courtesy G. O. Heintzen)

The stainless valves were improper, but if they slightly leaked during normal operation dry chlorine would enter the scrubber and harmlessly be neutralized. However, during this start-up with the piping under a vacuum, moisture from a slight leak did combine with the dry chlorine and create an aggressive corrosive fluid.

Teflon®-lined valve bodies and Teflon®-lined plugs were originally installed. Some of the supervisors seem to remember that the original 3 inch (7.6 cm) valves were very stiff to operate. It is believed that well-meaning individuals requested easier operating valves and no one worried about reviewing the plant specifications which were not always up to date. Even if piping specifications were consulted, this 3 inch (7.6 cm) branch line on a 24 inch (61 cm) line was itself a plant modification that was not properly authorized.

Substitute piping material installed – accelerated corrosion/erosion results in a large fire

As was stated in Chapter 4, piping systems in many plants are generally more likely to be ignored than other equipment such as: furnaces, reactors, heat exchangers and pressure vessels. The Institution of Chemical Engineers (UK) understands the problems that can occur in poorly designed, poorly operated or poorly maintained systems. The Institution offers an awareness training package entitled *Safer Piping – Awareness Training for the Process Industry* (Institution of Chemical Engineers, Hazard Workshop Module 12). This next example is a brief sketch of a very detailed piping failure and fire incident in this package.

In August 1984, Coker 8-2 was operating smoothly in the refinery until a piece of piping failed and a large fire broke out. The magnitude of the fire was so intense that it burned out of control for 2 hours, and it took another 2 hours to extinguish it. Over 2500 barrels of liquid hydrocarbon were lost during the fire (Institution of Chemical Engineers, Hazard Workshop Module 12).

A section of 150 mm (6 inch) diameter carbon steel fluid coker piping which was 450 mm (18 inches) long suddenly ruptured without warning. The gross failure of this slurry recycle line allowed a high initial release rate of hydrocarbons. A vapour cloud formed engulfing an area which included the light oil product pumps.

Within a minute the cloud was ignited. The rapid spread of the fire delayed containment and isolation. The unit suffered extensive physical damage, but there were no personal injuries (Institution of Chemical Engineers, Hazard Workshop Module 12).

Investigators determined that a carbon steel section of 150 mm line was installed in an area in which specifications required corrosion resistant 5% chromium/0.5% molybdenum (generally called 5-chrome) alloy piping. In a process plant like a fluid coker, the materials used are a mixture of carbon steel and other steel alloys. The welder and the maintenance crew, who previously installed this piping as a repair some time ago, were not aware of the piping specifications. They did not understand that if a material like carbon steel was installed in an area requiring 5-chrome alloy piping erosion/corrosion could cause failure.

After this expensive incident, the company developed and implemented a safety management system (Institution of Chemical Engineers, Hazard Workshop Module 12) to:

1. Provide positive material identification and to check the quality of *all* incoming material used for pressure containment systems in this process.
2. Involve the plant inspection group to witness and recommend procedures in all repairs such as major piping replacements in safety critical containment systems. (Their safety critical systems included high temperature, high pressure and highly corrosive applications.)
3. Provide ongoing training to ensure that plant personnel were aware of the need for and understood the appropriate standards and specifications.

Pump repairs endanger the plant – but are corrected in time to prevent newspaper headlines

There are many pumps within the typical chemical plant. Pumps require considerable maintenance and modifications can be introduced accidentally.

One type of misguided mini-modification (within a flammable liquids processing plant) was made in the name of employee industrial hygiene or safety. As a result of the US regulations phasing out asbestos gaskets, one plant replaced a number of asbestos pump bowl gaskets with low melting point non-asbestos gaskets that destroyed the 'fire safe' engineering of the pump design. Once identified as a problem, the oversight was quickly corrected.

Normal maintenance on a brine pump

The impeller within a large sodium chloride brine pump in a closed loop system of a chlorine plant needed to be replaced. The job was not assigned to the regular maintenance crew, but was was given to a crew which worked throughout a major chemical plant in areas that had a large backlog. This pump was specified to have all wetted surfaces fabricated of titanium. However, the job was being manned with 'outsiders' on an evening shift.

After the maintenance foreman searched for the appropriate titanium impeller and could not find one, he decided to improvise. The maintenance foreman located a stainless steel impeller which was available for other pumps of the same size and manufacture.

The job was completed and the maintenance foreman went to eat an overtime supper at the plant cafeteria. The maintenance foreman just happened to mention his ingenuity to the operating foreman, who was also working overtime.

After listening intently, the operating foreman explained that just traces of chrome salts in the brine system could create an explosive situation within the electrolytic chlorine cells. Traces of chrome salts in the feed brine to the chlorine cells liberate hydrogen gas in the chlorine cell gas. Hydrogen in the chlorine cell gas has a very wide explosive range. The use of stainless steel equipment in sodium chloride brine systems has devastated chlorine-processing equipment within other similar chlorine-manufacturing plants. The improper pump impeller was removed quickly before any problems occurred.

A cast iron pump bowl is installed in the wrong service

A flashing flammable chemical (that possessed flammability properties similar to those of butane) had been manufactured in this new unit for just a few years. The liquid phase reactor had a normal liquid inventory of 13 000 gallons (59 000 litres). The reactor was about 9 ft (2.74 m) in diameter and about 35 ft (10.7 m) high and operated at about 120 psig (826 kPa gauge).

Heat from this liquid phase exothermic reaction was removed by circulation of a side stream of crude product through large coolers. Reactor circulating pumps were designed with a 10 inch (25.4 cm) suction and an 8 inch (20.4 cm) discharge nozzle. These pumps continuously operated to cool the crude product and to help mix the incoming gaseous raw materials.

After several years of operation, conditions warranted the replacement of a pump bowl. Spare ductile iron pump bowls were not stocked within this plant. The maintenance force realized that there were a few other large pumps within the plant and they located one.

Unfortunately, they chose a cast iron pump bowl, which is very desirable for certain sodium chloride brine services, but totally unsuited for large volumes of highly flammable materials. Cast iron is undesirable (and prohibited by certain flammable liquid codes) because a brittle fracture can lead to a massive leak. Some months after this mistake, the unit received a loss prevention audit and as a result the pump bowls were replaced with ductile iron units. This same audit prompted the installation of remotely operated emergency shut-off valves around these pumps.

Plastic pumps installed to pump flammable liquids

In the late 1980s pneumatically-driven diaphragm pumps were gaining wide acceptance in handling acid-contaminated water streams. These plastic pumps could be installed by sumps and handle a wide variety of conditions.

These plastic diaphragm pumps had the advantages of:

1. Operating on compressed air or compressed nitrogen and having no need for electrical supply
2. Providing good suction pressure and being good for low flow conditions
3. Being seal-less as there was no rotating shaft
4. Being lightweight
5. Being compact with no pump/motor alignment problems
6. Being capable of running dry

With all of these advantages, it was easy for maintenance engineers to consider these pumps for many applications. Some of these pumps were installed to transfer flammable liquids in several units of a chemical complex.

These polytetrafluoroethylene (PTFE) pumps, along with the associated plastic expansion joints, were installed in the plant because no process safety engineer went to the plant site as modifications were being made. No one took the time to question their suitability in handling flammable liquids. They are just not 'fire-safe.'

Weak walls wanted – but alternative attachments contributed to the damage

Compressors handling flammable gases have the potential to leak. Ideally, such compressors should be installed outdoors so that leaks can disperse. However, at times, such as in very cold climates, compressors are designed to operate within buildings.

As a loss prevention feature, some flammable gas compressor buildings are designed with lightweight walls. If an explosion occurs in such a building, the lightweight wall panels, which are secured with break-away fasteners, will blow off as soon as a pressure rise starts. Proper design will prevent a higher, more damaging peak-pressure wave (Kletz, 1990b).

In one reported case, a construction manager reviewed the drawings of a gas compressor building before construction. He observed that very weak clips were specified to secure the plastic sheets to the structure. He apparently did not understand the concept, because he decided that they were too weak. He replaced the specified clip with a more robust clip.

An explosion occurred. The walls did not blow off until the pressure was much higher than the designers intended. The resulting damage was more severe than it should have been (Kletz, 1990b).

An explosion could have been avoided if gasket specifications were utilized

An ⅛ inch thick compressed asbestos gasket failed on a 20 inch (50.8 cm) class 600 ASNI flange, after a regular semi-annual turnaround. An explosion of about 30 kg (66 lb) of hydrogen that leaked from the manway flange on a compressor knock-out drum in a Canadian styrene plant resulted in two employee deaths and two more employee injuries (Kletz, 1990b).

In April 1984, the operating crew was attempting to restart the unit. They were trying to resolve problems so that they could increase the pressure on the hydrogen circulating system. A little known drain line modification was found to have an open valve. The operator reset the pressure controller to 720 psig (5000 kPa gauge). A 'pop' was heard followed by a loud hiss as pressure was being relieved. Within 10–15 seconds ignition occurred, which is common with high pressure hydrogen leaks. The resulting explosion took the lives of a foreman and an operator.

Witnesses reported seeing a white flash followed by a large fireball. Major fires in the unit burned out within 3 minutes, with the exception of a benzene fire from a pipe flange opened by the blast. This benzene piping could not be safely isolated from a vessel and the fire continued for 8 hours (MacDiarmid and North, 1989).

The explosion resulted in construction trailers having their sides bowed in, and significant window damages at a distance of 365 m (1200 ft) and as far away as 900 m (3000 ft). The explosion occurred on a plant holiday. If it had been a regular work day many more injuries could have occurred within the unit and in a neighbouring office building. This office building, which was located 120–150 m (400–500 ft) from the epicentre of the blast, sustained extensive damage to windows and minor structural damage.

Investigations revealed that the specified gasket for this installation was a spiral-wound asbestos-filled gasket with a centring ring. However, the maintenance personnel did not know the specifications, as they recalled that a compressed asbestos fibre (CAF) gasket had been used for 7 or 8 years.

Typically at this plant, substitution of a gasket is authorized only after calculations ensure that the substitution is acceptable. No calculations or authorizations were found. The substitution of a CAF gasket was probably intended as a one-time substitution and it was replaced in kind on each turnaround that followed. The ⅛ inch CAF gasket was theoretically acceptable in this service, but conditions would have required a full pressure test to ensure that the 24 bolts had snugged the gasket properly. Unfortunately, the risk of the gasket disintegrating under a high-pressure leak was not appreciated.

Another unauthorized modification contributed to the severity of the explosion, because well meaning individuals did not understand the design specifications of the compressor shed. The compressor shed was originally designed as a roof over the equipment. Later gables and a wall were added, perhaps for increased comfort during the winter operations and maintenance. This additional degree of confinement served to reduce the dispersion rate of the hydrogen and to contribute to the energy of the explosion (MacDiarmid and North, 1989).

Surprises within packaged units

Sometimes equipment which is rented or purchased as a packaged unit allows 'out-of-spec' components to enter a chemical plant. Air compressor systems, refrigeration units and boiler systems are prime examples of units which if improperly specified may allow out-of-spec elements into a refinery or chemical plant.

Caustic soda plants generally prohibited the use of any aluminium parts on any processing equipment, motors, instruments, etc., because aluminium is aggressively attacked by any fumes or drips of caustic soda. However, during a major expansion in the late 1960s, refrigeration units were purchased that were equipped with safety relief valves made with aluminium bodies.

During this very active period of construction the engineers of the project were relatively new and failed to consider the materials aspect. Several years later the aluminium safety relief valves were tested and were found to be corroded shut.

Another example of out-of-spec equipment is the use of filters on a unit. Many packaged units are equipped with filter cartridge bodies which have zero corrosion allowance and are not as robust as other equipment purchased for the process.

References

American Petroleum Institute (1979) *Safety Digest of Lessons Learned*, Section 2, API, Washington D.C., pp. 45–46
Griffin, M.L. (1989) Maintaining plant safety through effective process change control. 23rd Annual Loss Prevention Symposium, American Institution of Chemical Engineers, Houston
Institution of Chemical Engineers. *Fires and Explosions*, Hazard Workshop Module 003. Available as a training kit with 45 slides, booklets, teaching guides, etc. from the Institution of Chemical Engineers, 165–171 Railway Terrace, Rugby CV21 3HQ, Warwickshire, UK.

Institution of Chemical Engineers. *Furnace Fires and Explosions*, Hazard Workshop Module 005. Available as a training kit with twenty-three 35 mm slides, booklets, teaching guides, etc. from the Institution of Chemical Engineers, 165–171 Railway Terrace, Rugby CV21 3HQ, Warwickshire, UK

Institution of Chemical Engineers. *Preventing Emergencies in the Process Industries*, Vol. 1, Hazard Workshop Module 006. Available as a training kit with 60 slides, a video tape, booklets, guides, etc. from the Institution of Chemical Engineers, 165–171 Railway Terrace, Rugby CV21 3HQ, Warwickshire, UK

Institution of Chemical Engineers. *Safer Piping – Awareness Training for the Process Industries*, Vol. 1, Hazard Workshop Module 12. Available as a training kit with 105 slides, a video tape, teaching guide, etc. from the Institution of Chemical Engineers, 165–171 Railway Terrace, Rugby CV21 3HQ, Warwickshire, UK

Institution of Chemical Engineers (1981a) *Loss Prevention Bulletin, Articles and Case Histories from Process Industries throughout the World,* **32**, 13–14

Institution of Chemical Engineers (1981b) *Loss Prevention Bulletin, Articles and Case Histories from Process Industries throughout the World,* **32**, 2–5

Kletz, T.A. (1990a) *Critical Aspects of Safety and Loss Prevention*, Butterworths, London, p. 183

Kletz, T,A. (1990b) *Criticial Aspects of Safety and Loss Prevention*, Butterworths, London, pp. 122, 123, 183

MacDiarmid, J.A. and North, G.J.T. (1989) Lessons learned from a hydrogen explosion in a process unit. *Plant/Operations Progress* (April), 96–99

Sanders, R. E. (1983) *Plant Modifications: Troubles and Treatment*, Chemical Engineering Progress, pp. 76–77

Sanders, R.E., Haines, D.L. and Wood, J.H. (1990) Stop tank abuse. *Plant/Operations Progress* (January), 61–65

Chapter 7

'Imagine if' modifications and practical problem solving

'Imagine if' modifications – do not over-exaggerate the dangers as you perform safety studies

Jeff L. Joseck and Trevor A. Kletz created the material within the first part of this chapter. Jeff is a staff engineer employed by a major chemical plant in West Virginia (US); he has a solid background in loss prevention. Trevor Kletz is the most published loss prevention engineer of all time, with over seven books and over 100 technical articles. His topics in this chapter assume some of the modern synthetic materials which we use daily have been available since the dawn of time and that some ordinary materials are being examined in a sort of hazards review session. These are Kletz's most classic, clever illustrations of how we can sometimes be overzealous and too narrow in our focus on a problem.

Jeff Joseck's article was written to spark interest and discussion of process safety in safety meetings. It is also very skilfully worded to spark the reader into thinking about acceptable risks and minimal risks.

All three of these short articles show that it is easy to over-exaggerate the dangers of unfamiliar conditions and it is also easy for us to forget that people have learned to live comfortably with many of the hazards that surround them. The author thanks both of these writers for their permission to freely copy their work.

This first article was originally published in 1976 in the Imperial Chemical Industries Limited (ICI) Safety Newsletter. It later appeared in *Critical Aspects of Safety and Loss Prevention* (Kletz, 1990). It assumes that water, in the pure form, has been unknown – there are no seas, no rivers, no lakes – and it has just been discovered.

New fire-fighting agent meets opposition – 'could kill people as well as fires'

ICI has announced the discovery of a new fire-fighting agent to add to their existing range. Known as WATER (Wonderful and Total Extinguishing Resource), it augments, rather than replaces, existing agents such as dry powder and BCF (or Halon in the US) which have been in use from time immemorial. It is particularly suited for dealing with fires in buildings, timber yards and warehouses. Though required in large quantities, it is fairly cheap to produce and it is intended that quantities of about a million gallons should be stored in urban areas and near other installations of high risk ready for immediate use. BCF (Halon) and dry powder are usually stored under pressure, but WATER will be stored in open ponds and reservoirs and conveyed to the scene of the fire by hoses and portable pumps.

ICI's new proposals are already encountering strong opposition from safety and environmental groups. Professor Connie Barrinner has pointed out that, if anyone immersed their head in a bucket of WATER it would prove fatal in as little as 3 minutes. Each of ICI's proposed reservoirs will contain enough WATER to fill half a million two-gallon buckets. Each bucket-full could be used a hundred times so there is enough WATER in *one* reservoir to kill the entire population of the United Kingdom. Risk of this size, said Professor Barrinner, should not be allowed, whatever the gain. If WATER were to get out of control the results of Flixborough or Seveso would pale into insignificance by comparison. What use was a fire-fighting agent that could kill people as well as fires?

A Local Authority said that it would strongly oppose planning permission for construction of a WATER reservoir in this area unless the most stringent precautions were followed. Open ponds were certainly not acceptable. What would prevent people falling in them? What would prevent the contents from leaking out? At the very least the WATER would have to be contained in a steel pressure vessel surrounded by a leak-proof concrete wall.

A spokesman from the Fire Brigades said he did not see the need for the new agent. Dry powder and BCF (Halon) could cope with most fires. The new agent would bring with it risks, particularly to firemen, greater than any possible gain. Did we know what would happen to this new medium when it was exposed to intense heat? It had been reported that WATER was a constituent of beer. Did this mean that firemen would be intoxicated by the fumes?

The Friends of the World said that they had obtained a sample of WATER and found it caused clothes to shrink. If it did this to cotton, what would it do to people? In the House of Commons yesterday, the Home Secretary was asked if he would prohibit the sale of this lethal new material. The Home Secretary replied that, as it was clearly a major hazard, Local Authorities would have to take advice from the Health and Safety Executive before giving planning permission. A full investigation was needed and the Major Hazards Group would be asked to report.

Water is being marketed in France under the name EAU (Element Anti-feu Universel).

A process safety management quiz

Jeff Joseck is a staff engineer with Union Carbide; he developed this quiz and kindly offered it for publication. It stimulates the process safety engineer or the loss prevention engineer to think about acceptable and minimal risks. A few minor changes were made in Jeff's quiz to better fit the style of this chapter.

A process safety management problem is presented to the committee
Consider the following operation from the aspect of process safety. There is urgency to conduct this operation in the plant tomorrow.

1. Basically, it is a transfer of flammable liquid to a tank. The liquid will be 40–50°F with a flash point below −20°F. The fluid's electrical conductivity is low enough to create static.

2. The tank does not have an inert blanket. The tank is vented to the air and there was no flame arrester installed on the tank.
3. The inlet connection is above the liquid level throughout the transfer and it can not be easily altered.
4. The tank is steel, but does not have fixed grounding; grounding is expected to be made through the transfer line–hose connection.
5. The receiver tank's liquid level indicator will not be in service during the transfer. There is a separate, high liquid level switch that is interlocked to stop the transfer when the tank is full. The quantity transferred can be monitored from the distribution system, but it will not be specified.

Granted, these are not the safest conditions to make this transfer. But given these circumstances, do you think it would be an acceptable risk to proceed tomorrow?

Did the process safety committee reject this proposal?
If the committee said 'no', here is more information to consider:

1. The high-level interlock has proven to be reliable. There are claims it is functionally tested at least once a week.
2. The tank already contains a heel of the same material.
3. The transfer quantity is small (greater than 100 pounds but less than 1000 pounds), so it is more like drum filling than a tank transfer.
4. This tank has been filled with this material hundreds of times in the past, without incident.
5. The plant manager, a forceful practical engineer, knows of this operation and condones it.

Was this procedure judged to be an unacceptable risk?
If you said 'yes', what exceptions to these conditions do you make when filling your car at the service station? The 'transfer' described is actually the filling of a vehicle's petrol tank at the plant's pump.

The primary reason to oppose the 'filling of the car's petrol tank', as it was deceptively, but accurately, described, is simply that it violates the safe practices that have been adopted as standard. The initial question 'would it be an acceptable risk?' cannot be answered 'yes' based solely on the information given. One would need either to understand the physics, design standards, and routine testing that make 'fill-ups' safe, or to rely on history, which indicates low probability of incident. Without this knowledge, one would have to be conservative and say that petrol tank fill-up, as described, is not an acceptable, safe practice.

The purpose of this quiz
This quiz was not intended to remotely suggest that plants discontinue inerting tanks containing flammable liquids. It was not to suggest that we re-evaluate the safety precautions in our procedures and design criteria; on the contrary, this exercise emphasizes their importance.

Process safety professionals must be adamant in their responsibility to understand and control process hazards to the best of their abilities. However, such process safety individuals have a second responsibility, to recognize the difference between acceptable and minimal risk. The difference is a matter of ethics and economics.

New fibre production methods questioned

This clever article was composed by Trevor Kletz and appeared in an ICI Safety Newsletter in December 1978.

New fibre runs into trouble – 'land-use excessive, waste disposal impossible'
The proposals by ICI and other chemical companies to invest millions of pounds in the production of a new fibre are already meeting widespread opposition. While nylon and polyester and the other fibres which have been in use since the dawn of civilization are manufactured in conventional chemical plants, the new fibre, known as WOOL (Wildlife Origin Oily Ligament), will be grown on the backs of a specially developed breed of *Ovis musimon*.

Opposition to the new fibre centres on the extensive areas of land required for its production. While a million pounds per year of nylon or polyester can be produced in a fraction of an acre, the same quantity of WOOL will require at least 25 000 acres of good land or a larger area of hill land. This land will no longer be available for growing crops or for recreation.

For once the National Farmers' Union and the Ramblers' Association have combined to oppose a development and a public enquiry will be necessary.

The RSPCA has protested at the 'industrialization' of animals and has asked what will happen if they break loose from their enclosures. Although *Ovis musimon* is docile, all animal species, even man, produce occasional aggressive individuals.

Meanwhile the garment industry has pointed out the importance of quality control and has questioned whether the necessary consistency can be obtained in a so-called 'natural' product.

It is assumed that chemical plant process workers will operate the production facilities as they will replace plants traditionally operated by them. Assuming that operators will not be expected to walk more than 200 yards from the control room, the control rooms will have to be spaced 400 yards apart that is, one per 33 acres. Over 750 control rooms will be required for a million pounds per year operation. Building costs will therefore be much higher than on a conventional plant, and may well make the new process uneconomic, especially now that control rooms are being made stronger than in the past.

The greatest opposition to the new fibre is the result of the waste disposal problems it will produce. Vast quantities of excreta will be deposited by the animals and will presumably have to be collected and dumped. Have the risks to health been fully considered? Will decomposition produce methane and a risk of explosions? What will happen when the animals become too old for productive use? It has been suggested that they might be used for human food and it is claimed that they are quite palatable after roasting. To quote the Director of the Centre for the Study of Strategic Perceptions, 'The suggestion is nauseating. Five thousand years after the dawn of civilisation and 200 years after the industrial revolution, we are asked to eat the by-products of industrial production.'

Practical problem solving

Clever approaches to problem solving

When trying to find the 'best' approach for reviewing modifications, do not hesitate to seek out the most practical approach, even if others around you are choosing more academic and laborious procedures. For example, some engineers have laboured for days developing fault trees, and estimating failure rates to determine if a flame management system is an acceptable risk. Other more practical individuals reach for recommendations from the National Fire Protection Association and from industrial insurance companies for acceptable approaches for flame management systems. Some practical references are listed in Chapter 10.

In some other instances, engineers have counted flanges on long and complicated piping networks to help estimate potential leak sources. Their time would have been better spent getting an estimate on the maintenance mind-set, to determine what programmes were in progress to reduce leaks by the ever present threats of external corrosion.

The physics student and his mischievous methods

Mr. Alexander Calandra, who in 1968 was a member of the department of physics at Washington University in St Louis, Missouri, told a story of a physics student in a classroom setting. The student was mischievous and practical, but he was in an academic system that did not appreciate his abilities.

This anecdote was adapted from Alexander Calandra's article entitled 'Angels on a Pin' (Calandra, 1969) which was an excerpt from his book, *The Teaching of Elementary Science and Mathematics*, which was published in 1969. Calandra was asked by another instructor to help referee the disputed grading on an examination question. The physics teacher believed his student deserved a zero for his answer. The student was convinced he should receive a perfect grade, but felt the school was not interested in practical approaches to problems.

Previously, the instructor and student agreed that an impartial arbiter should review the examination question and Calandra was chosen. Calandra went to his colleague's office and read the test.

'How is it possible to determine the height of a tall building with the aid of a barometer?' was the disputed exam question.

The student's written response was 'Take the barometer and a long rope to the top of the building. Tie the rope to the barometer, and lower the barometer to the sidewalk. Measure the length of the rope that hung down the side of the building, and that is the building's height.'

Calandra argued with his colleague that the student provided an answer that was complete and correct, and therefore it seemed to the referee that the student was entitled to full credit. The troubled instructor cautioned Calandra that if he gave the student high marks for this answer, then the student was eligible for a high grade for the physics course.

The instructor was concerned that a high physics grade should mean the student performed commendably in the principles of physics, but this answer and several others like this one did not confirm that ability.

Calandra recommended to both the instructor and the student that the student be given another opportunity at answering the same question while in the instructors's office. Calandra advised the student that the answer must show some skill of physics and a complete answer must be written in 6 minutes.

At the end of 4 or 5 minutes the student still had a blank piece of paper. Calandra asked if the student wanted to give up. The student responded, 'No! I am just trying to decide which of my answers is the "best" one.'

The student quickly scribbled, 'Take the barometer and a stop watch to the top of the building and lean over the edge of the roof. Drop the barometer and time its fall with the stop watch. Then use the formula $S = \frac{1}{2} at^2$ to calculate the height of the building.' The arbiter showed the answer to his colleague and the student received almost full credit for this answer.

After the grade of the student was established, Calandra was curious enough to question the student about the other answers he was considering. The student replied, 'There are many ways to determine the height of a building with a barometer. For example, on a sunny day you can stand the barometer up and measure the shadow of the barometer and measure the shadow of the building. Using simple proportions the height can be calculated.'

The student volunteered, 'There is a very basic measurement procedure that would work very well. Set the barometer at the base of the stairs and draw a line at the top. Walk up the stairs and mark off the length of the barometer as you climb. Count the number of marks and multiply by the length of the barometer and that is the height of the building.'

The student continued, '. . . a more sophisticated approach would be to tie the barometer to a string. Swing it as a pendulum. Calculate the value of "g" at the street and again at the top of the building. From the difference between the two values of "g" the height of the building, at least in principle, could be calculated.' The student smiled and stated that he knew the conventional atmospheric pressure comparison answer, but that was neither the best practical answer nor the most accurate answer.

What is the best approach?

The student mischievously replied, 'The best and most accurate method to find out the height of the building is to take the barometer to the building superintendent's door. Knock on the door and say, "Mr Superintendent, I have this fine barometer to offer as a gift, if you will just tell me the exact height of the building." No answer can be better.'

In our world do we sometimes labour and struggle to determine a 'best' answer, when a simple call to the building superintendent (the resource person) would quickly provide accurate information?

References

Calandra, A. (1969) Angels on a pin. *AIChE Journal,* **15**, No. 2, 13. An excerpt from *The Teaching of Elementary Science and Mathematics*, ACCE Reporter, Ballwin MO

Kletz, T.A. (1990) *Critical Aspects of Safety and Loss Prevention*, Butterworths, London, pp. 342–343

Chapter 8

Programme to address ageing in chemical plants

Introduction

Gradual change takes place as chemical plants age. The sometimes unfriendly environment (including combinations of moisture, corrosion, dirt, various forms of deterioration, inexperienced or insensitive supervision and unauthorized tampering) can modify and compromise the integrity of containment systems and protective systems. It is difficult to introduce a section on 'Protective programmes' any better than Ken Robertson, president of Exxon Chemical Americas, stated in a keynote address at the Chemical Manufacturer Association's Plant Inspection and Maintenance Forum in 1990. Robertson indicated that public expectations are increasing and there is less tolerance of oil and chemical spills and tragic plant safety incidents (Robertson, 1990).

The Exxon president reviewed some of his company's process safety management practices relating to maintenance:

'First, safety critical systems must be reliable. These systems control releases in the event of accidents. It's necessary to have a critical analyser, instrument and electrical system test program. This should consist of preventive maintenance and alarm and trip device testing for panel alarms, emergency isolation valves and other critical components.

Also, procedures must be in place to control defeating safety critical systems. Before taking these systems out of service for any length of time, there must be proper authority, communication and detailed contingency planning.

Regular, comprehensive inspections to ensure the safe condition of site equipment is another important consideration. There must be clear lines of responsibility for inspection and maintenance of crucial containment systems. A formal system must be in place for documenting recommendations and communicating them clearly and quickly to the appropriate managers in the organization.

We're doing away with our traditional maintenance mindset of using heroic measures to fix something ... our approach is to take ownership, to use predictive tools to get ahead of problems . . .'

In the past, some managers may have given mixed signals by praising actions taken to keep production units on-stream in the face of critical alarms being activated. All supervisors and managers must be careful not to praise actions which allow production to continue while in violation of recognized safe procedures.

It takes little effort to raise the pressure alarm setpoint or the liquid level alarm setting in a way which might compromise the safety of the unit. Therefore there must be an operating discipline that does not age. The system should be developed to discourage and prevent insensitive or inexperienced management from reducing the safety margin.

Integrity assurance of the containment system

This section assumes that the existing chemical processing plant, including its storage and shipping facilities, were designed and fabricated to plant specifications and appropriate codes and standards. It is also assumed that the craftsmen were qualified, repair materials were certified (where necessary) and documentation of tests and inspections were retained. It furthermore assumes that any later modifications were also designed and fabricated in accordance with the appropriate codes and plant standards.

'Appropriate codes' include 'ASME Pressure Vessel Code', 'API Codes for Atmospheric and Low-Pressure Storage Tanks', the 'ANSI Piping Codes', the National Fire Protection Association Codes, property insurance guidelines, and any local or federal requirements.

In a recent article entitled 'Plant integrity programs', Krisher (1987) states:

'For an installed plant to achieve the level of integrity intended, a quality assurance program is needed to insure that the fabrication of new equipment is carried out with good engineering practice and standards. A similar program is required to assure proper quality of maintenance and in-plant modification work.

To realize the intent of a proper engineering design in the construction and installation of equipment, it is important to assign an adequate number of people with proper training and experience. The inspection program should be structured to uncover problems that are then brought to the attention of management. And management must be committed to it firmly. In today's cost-conscious business environment, inspection frequently is considered a cost to be minimized, not a function required to achieve corporate goals. Inspection programs often have been put in place in response to a major disastrous incident that had high costs to the business enterprise. When programs are installed in such response to such problems, they are frequently scoped and focused to prevent recurrence of a specific problem rather than to prevent all serious and dangerous incidents across the plant. An inspection program often is looked at in terms of the expended funds: 'What specific failures have been prevented.' If the program was truly effective, none will be visible. One can imagine what might have happened in the absence of such a program, but this frequently is not considered persuasive.'

In short, in chemical processes handling potentially hazardous materials such as flammables, combustibles, toxic chemicals or carcinogenic or environmentally objectionable materials, a system must be in place to ensure the integrity of the piping and vessels. Normally, this is accomplished by a competent equipment inspection team whose function is to detect any loss in integrity of material containing the hazardous fluids. The program should be capable of evaluating carbon steel, alloys, specialty metals, plastics, ceramics, or elastomers and other liners before conditions deteriorate to the point at which there is a loss of containment. A properly managed and knowledgeable inspection team should be viewed as a cost-saving resource.

Corrosion under insulation

During the 1960s and 1970s many plant designers were not concerned with the potential of problems of corrosion under insulation. As a result, today's troublesome pitting or lamellar rusting of carbon steel, chloride stress corrosion cracking of austenitic stainless steels and other concealed metal loss can be found beneath insulation. This hidden metal loss may result in drips, spills or ruptures and result in environmental insults, releases of toxic materials or conditions that allow sufficient flammables to escape and result in fires or explosions.

Intruding water is the menace that encourages corrosion problems on insulated equipment. Very few, if any, thermal insulation systems are completely waterproof, and the steel or alloy material covered by insulation is intermittently wet and dry. Brackets which support insulated vessels or brackets on insulated vessels to support piping, ladders, or walkways can act as traps to collect water.

Special care must be taken during the design stages and over the life of metallic equipment to reduce or eliminate the intrusion of water into the insulation by direct openings or by capillary action. The weather/vapour jacket covering the insulation provides the primary barrier to rain, drips and areas which are washed down. This barrier can be destroyed by walking on piping, by poor maintenance, and by weathering.

In early designs, thermal insulation was applied over bare steel or one coat of oil base primer. According to the National Association of Corrosion Engineers (1989), the most serious corrosion problems seemed to be most prevalent in plants with chloride-containing or sulphide-containing environments. These corrosion problems were further aggravated in high rainfall, high humidity and salt air locations.

NACE studies determined that the most severe corrosion problems occurred under thermal insulation at temperatures between 140 and 250°F (60–120°C) where the temperatures were too low to quickly boil off any intruding water. At temperatures higher than the boiling point of water, other corrosion problems can occur. The intruding water can carry chlorides and other corrosive elements which can concentrate and result in stress corrosion cracking.

When serious corrosion was first noted, many engineers decided that better surface preparation and primer were needed. Inorganic zinc was then often selected and used to prime carbon steel which was to be covered with thermal insulation. In the 1970s some US Gulf Coast corrosion engineers noted that the presence of inorganic zinc seemed to accelerate corrosion.

A NACE task group (National Association of Corrosion Engineers, 1989) reported in 1986:

1. Inorganic zinc primers did not perform well under insulation, even if they were top-coated.
2. Catalysed epoxy coatings lacked sufficient high-temperature resistance to soaked insulation for the long-term exposure.
3. The best generic coatings for carbon steel which may be surrounded by wet insulation in the 140–250°F (60–120°C) range are the epoxy phenolic and the amine-cured coal tar epoxy formulations.
4. Chloride stress cracking corrosion was also a problem when austenitic stainless steels were covered with thermal insulation. The NACE paper also discusses the protective coatings for stainless steels and provides sandblasting and coating application guides for stainless and carbon steels.

To diminish the undesirable modifications by ageing within chemical plants, the design engineer, operator supervisor and maintenance foreman must focus on the problems of corrosion under wet insulation. Certain practices should be followed, including:

1. Providing an appropriate coating system for all new installations.
2. Providing a periodic inspection of the piping or equipment in areas around potential points of moisture intrusion by removal of insulation.
3. Repairing mechanical damage for weather barriers and thermal insulation.
4. Avoiding the use of zinc, cadmium, or zinc-rich coatings on austenitic stainless steels. Even stainless systems operating at low temperatures can accidentally be heated by fire or welding repairs and can cause stress cracking.
5. Becoming familiar with NACE Publication 6H189.

Inspecting pressure vessels, storage tanks and piping

In the typical chemical plant, inspections primarily address the corrosion and weld repairs of steels and alloys. However, the following elements also need regular inspections to ensure containment: the equipment's foundation; the connecting piping; exterior coatings; insulation; rubber linings or ceramic linings within reactors or tanks; glass fibre vessels and piping. The inspection function should be conducted and managed by personnel who are independent of production, not unduly influenced by limited maintenance budgets, and still have management's ear.

It is not the intent of this chapter to review in detail the types of inspection tools, techniques and procedures. Recent American Petroleum Institute publications are available at moderate costs which do an excellent job in that respect.

Inspection of pressure vessels and storage tanks

The inspection of pressure vessels is addressed in API 510 *Pressure Vessel Inspection Code: Maintenance, Inspection, Rating, and Alteration* (American Petroleum Institute, 1989). API 510 is an excellent reference. Some countries, provinces, and states require pressure vessels to be inspected by government agents. In those cases, the governing agency rules should be the standard reference.

External inspections should be made regularly to operating vessels to help ensure the integrity of the vessel. The frequency of this inspection should be between semi-annually and once every 3 years for most chemical plant processes, depending on the aggressiveness of the chemical plant atmosphere; however, API 510 will accept intervals up to once every 5 years.

Visual external inspections are often made with the normal physical limitations of accessibility and visibility that exist at the time of the inspection. Normally, the removal of insulation, and scaffolds, manlifts, etc., are not employed during frequent external inspections unless 'telltale' conditions warrant a closer examination.

The following items should be considered during an external inspection whenever applicable:
1. Anchor bolts – should be painted and free of corrosion; this is especially important on tall towers.
2. Foundations – should show no signs of serious spalling, cracking or settling.
3. Supports and skirts – should be free of corrosion and signs of deflection or buckling; ideally, skirts and supports should be accessible for inspection, sandblasting and painting.
4. Tank accessories – ladders, walkways, pressure gauges, sight glasses, associated piping (including supports, pipe hangers, etc.) should be serviceable. Ladder rungs, stair treads, and handrails should be carefully examined as the tank is being climbed.
5. Coatings – paint on metals should be intact to reduce corrosion potentials.
6. Grounding – grounding cables should be securely attached to the tank and the grounding grid in the foundation.
7. Heads and shell – these containment elements should be free of corrosion, distortion, obvious cracks, dents, bulges, blisters, signs of leakage, etc.
8. Low-pressure tanks – special attention should be given to the shell near the bottom on low-pressure tanks since corrosion is sometimes caused by corrosive soils or trapped water. Tank roofs also require detailed inspections to establish that they are safe to stand on.
9. Fireproofing or insulation – should be weather sealed, and intact to prevent collection and retention of moisture against metal surfaces which could corrode.
10. Nozzles, flanges and valves – should be free of signs of corrosion and leakage. All flange bolts should be installed and be of the correct diameter and length.
11. Relief devices – Safety relief valves should be mounted in the vertical

position. Combinations of rupture discs mounted below safety relief valves should have a pressure gauge or other device to warn that the rupture disc is leaking.

There is no reliable substitute for a detailed internal inspection (Figure 8.1). Internal inspections are essential to determine if there is any weakening of the vessel or any conditions which may develop into unwanted leaks. Typically, internal inspections are established with frequencies between semi-annually and once every 10 years. The exact frequency is best determined by the corrosive nature of the chemical being processed or stored, including the effect of trace components and the past history of this equipment.

The global view of a vessel provided by an internal inspection can be much more revealing than any of the limited judgements based upon thickness readings from the outside of the tank. Visual examination is usually the single most important and universally accepted method used to ensure fitness of service of a vessel. The visual inspection should include examination of the shell and heads for cracks, blisters, bulges, operating deposits, corrosion products, erosive patterns and other deterioration. In various aggressive chemical services, it is important to examine the welded joint and the heat-affected zone for accelerated corrosion, cracks and other metal loss. Often the use of a torch (US, flashlight) placed parallel to the internal walls of a vessel shining away from the inspector will help to visually identify metal loss, or distortion of the walls.

Another often-used test is metal thickness verification, using the simple hand-held ultrasonic thickness meter. Ultrasonic wall thickness measurements resemble radar or sonar in technique. A burst of ultrasound is emitted via a probe into a material to be bounced off the rear wall. The time interval measured for this reflection to return is a measure of wall thickness.

Figure 8.1 Metals inspector preparing to internally inspect a vessel

If the vessel is not insulated or does not have a liner, the wall thickness may be tested either from the outside or from within. Often corrosion patterns are not uniform, so it is best to check the thickness from the inside. Other inspection methods to ensure mechanical integrity of vessels may include: magnetic-particle examination for cracks at or near the metal surface; dye penetrate examination for disclosing cracks or pin holes that extend to the surface; radiographic examination; acoustic emission testing and pressure testing.

The inspection of low-pressure tanks is amply covered in API 653, *Tank Inspection, Repair, Alteration, and Reconstruction* (American Petroleum Institute, 1991). This short, practical booklet covers such items as inspector qualifications, concerns for tank bottom corrosion, etc. Appendix C offers a four-page checklist for inspection of a tank while in service and seven pages of checklists for internal inspections.

One chemical plant's pressure vessel management programme

Recently the Freeport, Texas (US) Dow chemical plant shared details of their pressure vessel management programme (Mueller *et al.*, 1986). The Dow Texas pressure vessel programme had evolved over 25 years and covered about 12 000 pressure vessels.

The scope of pressure vessels covered by Dow included any vessel with an internal diameter greater than 6 inches (15 cm) and having a relief valve setting of 18 psig (125 kPa gauge) or higher or a rupture disc with a bursting pressure of 25 psig (170 kPa gauge) or greater. The programme included vendor or contractor equipment, but excluded residential water heaters (Mueller *et al.*, 1986).

Vessel design review and registration
The Pressure Vessel Review Committee determines if a vessel meets the requirements of the ASME Code and/or Dow's engineering specification. If a pressure vessel meets those requirements, it is assigned a unique index number. Such a pressure vessel is either fabricated in a commercial fabrication shop, received as a part of a package unit (like a refrigeration unit or an air compressor skid) or built within a Dow fabrication shop (Mueller *et al.*, 1986).

If a vessel is to be fabricated outside of the Dow Chemical Company, the requisition must be routed to a Dow buyer who has attended training courses and is familiar with engineering specifications. This buyer works under the guidance of a Custom Fabrication Purchasing Agent. Procedures require that all specifications, drawings and requisitions are approved by a pressure vessel 'designer' who is also responsible for reviewing vendor calculations and fabrication drawings.

These procedures ensure that each vessel is reviewed by at least two Dow employees, plus employees of the fabrication shop, prior to the start of construction. A pressure vessel inspector follows the fabrication and ensures that the company specifications are met and that ASME Code considerations are achieved. The inspector also arranges to have a set of wall thickness measurements made to be a baseline for any corrosion studies. Therefore, before a vessel is put in service at least three Dow employees have examined various aspects of the quality of the job (Mueller *et al.*, 1986).

The Pressure Vessel Review Section (PVRS) is a service organization that assigns index numbers, and maintains current records and inspection status on all vessels. The PVRS files include:

1. Manufacturer's data report
2. Registration form
3. Actual thickness measurement reports
4. Minimum allowable thickness calculations
5. Relief system design calculations

In addition, the typical owner files include:

1. Vendor calculations
2. Mill test reports
3. Shop drawings
4. Visual inspection reports

Vessel inspections at Dow
The Dow Freeport inspection programme was reported to include an annual visual inspection while the vessel is in operation. The initial out-of-service (or internal) inspection is scheduled within the first 5 years a vessel is in service and subsequent internal inspections are performed at least every 5 years thereafter. If either the owner or the inspector feel it is necessary for more frequent in-service and out-of-service inspections, the schedule is adjusted to accommodate that need. The PVRS can extend the frequency of internal inspections up to once in 10 years, if sufficient data support this conclusion (Mueller *et al.*, 1986).

The Dow Freeport in-service inspection procedures are similar to those reported earlier in this chapter. The Dow out-of-service (internal) inspection includes ultrasonic thickness measurements at all benchmark locations. Other test methods include shear wave ultrasonics, eddy current and radiography. The ultrasonic thickness readings are used to project the remaining useful life of the vessel and when the next internal inspection should be scheduled.

Repairs, revisions, and other status changes on vessels
Under this system, Dow equipment owners are required to report a change of status when any of the following occur (Mueller *et al.*, 1986):

1. Rerating of the vessel
2. Change in emergency relief requirements
3. Change in service
4. Return to service after being idle for a prolonged period
5. Repair or modification to a pressure-containing part

If pressure-containing parts of a vessel corrode or erode to a thickness less than the minimum allowable, then it must be reconditioned or given a lower pressure consistent with the new conditions. When a vessel is rerated, the relief system is reviewed (Mueller *et al.*, 1986). The Dow system seems to be a very practical method to acquire and maintain pressure vessels in a safe manner.

Inspection of above-ground piping

The inspection of piping is very well covered in the API 574, *Inspection of Piping, Tubing, Valves, and Fittings* (American Petroleum Institute 1990). The recommended practice addresses the description of piping components, causes of deterioration, frequency of inspection, inspection tools, inspection procedures and record keeping.

The guide is full of practical information including piping inspection procedures, with these hints:

1. Inspect areas in which the piping could have experienced metal loss because it was affected by velocity or turbulence. This includes the outer surfaces of piping bends and elbows, areas around tees, piping around reducers, orifice plates, and throttling valves and the piping downstream from them.
2. Inspect areas of piping at which condensation or boiling of water or acids can occur.
3. Pay attention to slurry piping, and piping in which catalyst or other abrasive materials are conveyed.
4. Deadends subject to turbulence, condensation or areas in which freezing is likely to occur should also be examined.
5. Piping areas just downstream of a corrosive chemical injection point should receive particular examination.
6. In many cases corrosion under piping insulation is a substantial threat. Corrosion is found adjacent to pipe supports, on small-bore piping for the vents or relief valves as the small diameter piping protrudes from the top of an insulated larger line or protrudes beneath the pipeline. Water can enter and accumulate in these locations.

Corrosion under insulation, as generally discussed above, is a genuine concern, because it can be so easily concealed. One plant reported that much of its piping had to be replaced after a major expansion required removing insulation that had been covering piping for over 20 years. Some 8 inch piping with 0.375 inch (0.95 cm) wall thickness which operated at 475 psig (3300 kPa gauge) had pitting so deep that a mechanic could push his finger through the pipe wall. Also, a 6 inch pipe with 0.375 inch (0.95 cm) walls and which operated at 500 psig (3450 kPa gauge) had external pitting 2 inches (5 cm) in diameter and 0.160 inches (0.41 cm) deep. There were also many small-bore pipes with pinholes corroded through in many areas (National Petroleum Refiners Association, 1990).

The *'Safer Piping: Awareness Training for the Process Industries'* (Institution of Chemical Engineers, Hazard Workshop Module 12) discusses a survey of 13-year-old carbon steel piping. The thermally insulated piping that was examined carried liquids at temperatures between $-10°C$ and $140°C$. Initially, about a third of the pipes in a 10 km (6.2 mile) network were inspected and the following was observed:

1. 10% of the piping had a 50% wall thickness decrease.
2. 10% of the piping had a 25–50% wall thickness decrease.

3. 30% of the piping had a 10–25% wall thickness decrease.
4. 50% of the piping had a 10% wall thickness decrease.

The Hazard Workshop Module 12 module sums up the corrosion on piping under insulation as a problem of water ingress on piping with poor surface protection. Engineering, construction and maintenance recommendations parallel those presented in 'Corrosion under insulation'.

An assurance program for safety relief valves and safety critical instruments

Safety relief valves (SRVs) and safety critical alarms and critical shutdown systems can be modified easily by ageing or tampering. The local environment surrounding the device, including cycles of freezing and thawing, moisture, corrosive contact with equipment internals, localized corrosive emissions, general atmospheric corrosion, dirt, sand-blasting, painting and tampering, can alter the ability of these process safeguards to properly function.

McGraw-Hill Publications has granted special permission to use excerpts from a previous PPG Industries article in *Chemical Engineering* entitled 'Don't leave plant safety to chance' Volume 98, No. 2, February 1991 (Sanders and Wood, 1991). Many paragraphs, like the next three, will be used verbatim in the rest of this section on SRVs without being shown in quotes.

All of the effort expended in designing plant-safety systems is of little value unless accompanied by an adequate 'proof-test' programme and regular maintenance. These safety systems – consisting of such components as safety-relief valves, tank vents, critical alarms and protective isolation and shutdown devices – do not operate on a continuous basis. Rather, they are only called into service periodically, to warn of, or to prevent, conditions that could lead to plant accidents (Sanders and Wood, 1991).

After just a few years of neglect, many protective devices and preventive process loops could become ineffective. Unfortunately, these safety-system failures may go undetected until a crisis occurs.

Determining the methods and frequencies of testing for critical instrument loops must be tailored to each plant's needs and resources. No matter how often testing is carried out, the facility must be committed to checking all plant safety devices regularly. Testing programs must be impervious to all the other demands taxing a plant's resources, including key personnel changes. Periodic in-house audits should be conducted to ensure that testing methods are effective.

Safety relief valve considerations

Safety relief valves are incorporated into the overall process safety of a chemical plant. The decisive functions of these safeguards are described in an article in *Valve Magazine* (Braun, 1989) which begins with this attention-getting paragraph:

'Perhaps no one valve plays as critical a role in the prevention of industrial accidents as the pressure relief valve. This "silent sentinel" of industry, sometimes referred to as the "safety" or "safety relief" valve, is essential in helping us minimize industrial accidents caused by the over-pressurization of boilers and pressure vessels.'

The following observations were reported by Van Boskirk (1986) at an AIChE Meeting in Southeast Texas:

'Relief valves are very deceiving in appearance. Because they look like pipe fittings, they are often handled and stored in the same manner. This misconception can lead to significant abuse and damage. RELIEF DEVICES ARE DELICATE INSTRUMENTS AND MUST BE TREATED AS SUCH.

It is reasonable to draw an analogy between a shotgun and a relief valve. Both are made from high quality materials. Both contain parts machined to extremely close tolerances. Both are expected to operate accurately and reliably when called upon. Both are intended to provide many years of dependable service. AND BOTH CAN BE EASILY RUINED THROUGH ABUSE, MISUSE AND NEGLECT.'

'In-house' testing safety relief valves

Undetected corrosion, fouling of safety relief valves (SRVs) or fouling in the inlet or outlet piping of the SRV can adversely affect the set point and the relieving flow rate. Equipment that has been improperly specified, installed or maintained is particularly susceptible to undetected corrosion and fouling.

One major chemical complex, PPG Industries, Lake Charles, LA (US) relies on about 2200 individual SRVs and conservation (pressure/vacuum) vents for overpressure protection. This plant has a number of unique materials and corrosion challenges. On a typical day, huge quantities of saturated brine (NaCl) enter the complex, and several thousand tons of chlorine are produced. In the humidity of Southern Louisiana, the salt, chlorine, co-product caustic soda, and hydrogen chloride gas circulating throughout the plant can attack process equipment. Other products such as vinyl chloride monomer, methyl chloroform and other chlorinated solvents require unique elastomers for 'O'-rings and gaskets.

PPG Lake Charles was one of the first chemical plants to openly discuss problems that users experience by improperly specifying and testing SRVs. Expertise on new safety valves could be obtained from manufacturers; sizing methods were documented by various designers and users, but no one published practical SRV testing and SRV corrosion problems.

Over the years, PPG Lake Charles has shared its findings with a number of organizations, including the Louisiana Loss Prevention Association, the American Institute of Chemical Engineers, and the Chemical Manufacturers' Association. A series of technical papers (Sanders and Wood, 1991; Sanders and Woolfolk, 1984, 1987) has generated considerable interest. As a result, PPG has also invited representatives of over 20 different chemical, petrochemical, and petroleum companies to see their test facilities and share information.

Warren Woolfolk, Inspector, was assigned to be PPG's first SRV Test Program Coordinator in 1974. His responsibility was to ensure that SRVs were properly specified, installed and tested, and that there was sufficient record keeping. He gathered information from SRV manufacturers, helped develop centralized record keeping, and improved engineering specifications.

Woolfolk became concerned that the SRV test facilities, which were commonly used by many commercial SRV test and repair shops and most chemical plants, were inadequate. The typical test stand consisted of a clampdown table connected to a compressed air cylinder by a ¼ inch (0.6 cm) tubing. Warren Woolfolk experimented with surge accumulators to test relief valves, using a substantial volume of air to mimic actual process conditions.

In 1983, after some years of testing SRVs with sufficient volume of gas, Woolfolk presented 'Process safety relief valve testing' (Sanders and Woolfolk, 1984) to the Loss Prevention Symposium of the American Institute of Chemical Engineers. The resulting published article created lots of interest; Woolfolk presented an updated version to the Chemical Manufacturing Association in 1986. This second article was published in *Chemical Engineering* (Sanders and Woolfolk, 1987).

However, a number of US chemical plants were still using the primitive clampdown table connected to a compressed air cylinder in the late 1980s. (See Figure 8.2.) The American Society of Mechanical Engineers (ASME) understands the need for a sufficient volume of gas to properly test a safety relief valve. An ASME subcommittee is currently shaping a policy statement on the optimum volume accumulators to test SRVs.

Selection of safety relief valve testing equipment

Until the ASME provides some defined guides, a concerned chemical plant may wish to develop its own criteria for volume testing or gather current data from various SRV manufacturers. Limited-volume testing often fails to result in a distinct 'pop'. (A 'pop' is defined as the rapid opening of the SRV with its associated audible report.) If a valve only 'simmers' instead of making an actual 'pop', internal parts can misalign on some SRVs, resulting in leakage (Sanders and Woolfolk, 1987).

The simmer point is seldom at the same pressure as the 'pop' pressure. Most spring-loaded SRVs simmer at 90–95% of the set or popping pressure. In other words, some SRVs could be set 5–10% too high (and in cases where springs were faulty larger errors of up to 20% or more could occur). Simmering does not guarantee that the SRV will fully open to discharge name plate flow capacity.

PPG Industries Lake Charles used several salvaged vessels as accumulators to test various SRVs. The objective was to optimize on the proper sized accumulator. An oversized accumulator could be costly in compressed air usage, while an undersized tank would not provide the necessary volume to properly respond.

After limited testing to estimate the required accumulator volume for their needs, a decision was made that a 3ft^3 (85 litre) test tank (or accumulator) rated at 500 psig (3500 kPa gauge) would provide a

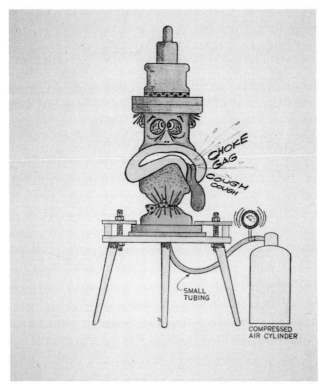

Figure 8.2 A cartoon of a safety relief valve being tested on a test-stand *without* sufficient volume (Courtesy of J. Fall)

representative pop action for an SRV with a normal blown-down ring setting up to and including a 3 inch 'L' orifice SRV. For larger SRVs up to and including a 'T' orifice valve, a 15 ft^3 (425 litre) test tank also rated at 500 psig was provided. For this plant's higher pressure SRVs, a 4.4 ft^3 (125 litre) tank with a 3000 psig (20 700 kPa gauge) rating was installed. A smaller tank ⅓ ft^3 (9.4 litre) rated at 5000 psig (35 000 kPa) that can be used pneumatically or hydraulically was provided for some special cases.

A few years later an additional 6.5 ft^3 (184 litre) tank engineered for vacuum and low-pressure applications was installed. Any plant considering installation their own test facilities should carefully study the specific needs of their unique situation.

All test accumulators were mounted directly below the test table. These vessels were mounted vertically with a 2:1 ellipsoidal heads on top and bottom. The top head has a well-rounded outlet nozzle sized equal to the SRV inlet to be tested. To test smaller-orifice SRVs, ring-shaped adaptor plates having an opening equal to the SRV inlet are used. This arrangement reduces the possibility of starving the flow to the valve.

The accumulator depressurization piping is connected to the centre of the lower head of the accumulators. This arrangement ensures that any

Figure 8.3 Safety relief valve test-stand for dynamic testing

moisture, pipe scale or other foreign material is expelled on each tank depressurization. For a complete description of this system see Sanders and Woolfolk (1987).

SRV testing and repair procedures

Most small chemical plants with just a few dozen SRVs must consider the local SRV test and repair shops for help. The larger chemical plants must decide if they want to invest in equipment and factory training for the mechanics, and approve the capital for an inventory of spare parts to properly maintain the SRVs.

If a chemical plant decides to operate an in-house SRV test and repair programme, it must have the necessary resources. These essential resources include sufficient volume test equipment, factory-trained mechanics, a supply of spare parts and an up-to-date quality control manual. Valve test performance should be carefully monitored under simulated process conditions to ensure that an adequate volume of gas (usually air) is being used.

A good quality control manual should include specific actions to accommodate testing requirements. The SRVs should be pre-tested and the appropriate data on the initial pop pressure and blowdown observed and recorded. This is the procedure for all valves removed from process service as well as for brand new valves (Sanders and Wood, 1991).

The safety valve is disassembled and examined, and all internal parts are blasted with glass shot, to remove rust and surface accumulations. The valve body is then sandblasted and repainted, both internally and externally.

During this stage, all defective parts are replaced or reconditioned to the manufacturer's specifications. Metal seats are machined to the desired

Figure 8.4 Well-trained mechanic testing a safety relief valve

flatness. All moving or guiding surfaces, and all pressure-bearing surfaces and parts, are lightly lubricated (Sanders and Woolfolk, 1987).

Once the valve is reassembled, the SRV is performance tested and adjusted to the correct set pressure, and the blowdown is set. All bellows-type and spring-loaded valves exposed to back pressure are back-pressure tested to ensure the integrity of the bellows and gaskets on the exhaust side of the SRV (Sanders and Woolfolk, 1987).

Tamper-proof seals are added to discourage unauthorized or undetected field adjustments. Once testing is complete, a sheet metal identification tag (usually of soft lead, which is easy to imprint, yet impervious to corrosion) is attached to the SRV. The tag records the location number, set pressure and test date.

Valve openings are sealed with tape and cardboard to prevent the entry of foreign material during transit and storage, and lifting levers are tied down to protect the SRVs' seats in transit to the user. The valve must be transported and stored in an upright position.

During testing, results are recorded on an audit form checklist. Information recorded for each SRV includes initial test pressure, and the condition of seats, stems, guides, springs, and inlet and outlet piping upon arrival at the plant. This information is sent to the appropriate area supervisor, and a copy is maintained in the central SRV records (Sanders and Wood, 1991).

How often is often enough when testing SRVs?

PPG Lake Charles has established different test cycles for different process safety equipment. Some are tested as frequently as every 6 months; some are tested as infrequently as once every 3 years. SRVs operating in dirty, fouling or highly corrosive conditions may require more frequent testing to ensure reliable service.

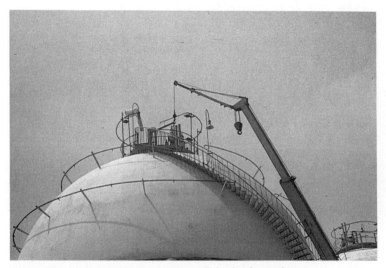

Figure 8.5 Preparing to remove a safety relief valve for testing

State laws, corporate policies, insurance regulations, and any other appropriate mandates should be considered when establishing regular testing cycles. Additionally, any SRV that pops or discharges a corrosive or tar-like fluid should be removed and reworked as soon as the discharge is detected. At the Lake Charles facility, about 30% of the overpressure devices are tested annually: another 60% are tested every 24 months and the remainder are tested at other frequencies. In general, SRVs on the following types of equipment are tested at the corresponding intervals (Sanders and Wood, 1991):

1. Positive-displacement pumps or compressors – every 12 months
2. Vessels processing corrosive chemicals – every 12 months
3. Processing vessels with heat sources, including stills, kettles, hot-oil systems and low-pressure, refrigerated storage tanks – every 12 months

4. Boilers – every 12 months (Louisiana law requires boiler SRVs to be tested, set and sealed annually)
5. Process steam headers – every 24 months
6. Storage vessels containing clean (non-plugging, non- corrosive) fluids – every 24 months
7. Instrument air manifolds within dehumidified control rooms – every 36 months
8. Lubricating relief valves on enclosed systems, such as compressors, turbines and generators – every 36 months

Keep the SRV records straight

A simple, sequential numbering system is acceptable for a small plant, such as a batch-chemical processing facility with only a few dozen SRVs. However, for the Lake Charles plant, a dual-numbering system is used to keep up with 2000 SRVs in service and over 600 additional SRVs in storage. The devices kept in storage on-site are used as spares, or can be cannibalized for parts. Spare safety valves for critical equipment must be available (Sanders and Wood, 1991).

In the dual-numbering system, one number refers to the valve type, and the other refers to its location and service. The first number, identifying the equipment, is unique to the SRV, and is retired once the valve is taken out of service.

Equipment records include information on the orifice size, set pressure, manufacturer, model number, serial number, and inlet and outlet flange sizes and ratings.

Test results are entered into the mainframe computer according to each device's identification number. Personnel have 'read' access to the records at over 500 terminals and personal computers throughout the complex.

The importance of proper identification of the SRV cannot be overstated. Without a strict numbering system, the risk of incorrectly joining the wrong SRV to the right vessel after testing is complete is increased.

For example, a 1 inch threaded valve can have an effective orifice area of from 0.06 to $0.314\,in^2$ (a five-fold difference), yet both could fit the same installation. Similarly, a 1 inch threaded-type SRV set at 100 psig (690 kPa) may look identical to another valve with the same description but set at 200 psig (1380 kPa).

Similarly, a 4 inch valve with 300 psi style flange could be equipped with either an 'L' orifice ($2.85\,in^2$) or a 'P' orifice ($6.38\,in^2$). The 'P' orifice is 2.24 times larger than the 'L' orifice (Sanders and Wood, 1991).

The location portion of the numbering system indicates the SRV's specific site. Site records include a description of the location, temperature, set pressure, product and presence or absence of a rupture disc.

Communications to equipment owners and management

Every other month, a computer-generated inspection schedule is sent to the supervisor of each process unit. The form serves as a reminder of all of the SRV and conservation vent locations, recommended service and frequency of inspection, last test date and results (satisfactory, changed or unsatisfactory), and the correct set pressure, back pressure, operating temperatures, and equipment number of each SRV and conservation vent. This schedule is intended to highlight those valves due for testing.

In addition, brief, statistical SRV test compliance summaries are prepared and these reports are circulated to higher levels of management. These compliance reports list the total number of SRVs in the area and the total number (and total per centage) that are overdue for testing. They serve as a periodic report card, signifying which process areas are in compliance, and which are delinquent (Sanders and Wood, 1991).

Process safety interlocks and alarms

As mentioned previously in this chapter, process safety interlocks and alarms can easily be modified by ageing or tampering and these essential components are just too vital to ignore. Each organization must develop an effective way to ensure that process safety systems will properly function when the process demands protection.

Whereas most SRVs share many similarities, process safety interlocks are each individually tailored to the unique requirements of the process. The final parts of this chapter will discuss the critical process shutdown systems and the process safety alarm management systems used by three different major chemical plants.

Protecting process safety interlocks at a DuPont plant

The Sabine River Works of the DuPont Company plant in Orange, Texas (US) shared their process interlock classification and test programme several years ago at a Chemical Manufacturers' Association Meeting, and it was later published by the American Institute of Chemical Engineers (Barclay, 1988). This large DuPont chemical complex consists of large single train processing units. Most of the units are continuous processes for the manufacture of alkenes, polyalkenes and chemical intermediates for nylon. There are about 2500 people working within this complex.

Process materials range from mild-mannered substances such as water and air to highly toxic chemicals and very large inventories of hydrocarbons. Process temperatures were reported to range from −180°F (−118°C) to 1000°F (540°C). Process pressures were reported to range between a low vacuum and over 30 000 psig (207 MKa) (Barclay, 1988).

The DuPont Sabine River Works has between 35 000 and 40 000 instrument installations and more than 5000 safety interlocks and alarms. The interlocks and alarms are divided into two classifications which are either an operator aid or a safety interlock alarm.

The *operator aid* is used to alert the operator of a non-hazardous abnormal condition that might otherwise be undetected. The *safety interlock or alarm* refers to any equipment whose proper functioning is essential to prevent or signal hazardous process conditions that may threaten personnel or equipment (Barclay, 1988).

DuPont's Sabine River Works' experience shows that it is crucial to segregate the more critical interlocks from the others. A few well-understood, respected interlocks are much better than an overwhelming number of interlocks with the critical ones scattered amongst the others. DuPont's system allows a range of required authorization; for example, a foreman may bypass less critical interlocks such as lubrication alarms for up to 24 hours for testing and repairs, but some of the more critical alarms may not be bypassed by anyone.

Testing safety critical process instruments at a DuPont plant

Interlock testing is performed on a periodic basis, usually every 12 months, following detailed written test procedures. The test procedure for each safety interlock defines the actions, including the approval system and the mechanical work required to ensure reliability. These procedures cover the testing activity from the process measurement device to the final element, which is often a control valve (Barclay, 1988).

Due in part to the complexity of the instrument systems, there are two first-to-final testing methods. These methods are called 'actual tests' and 'simulated tests'.

The 'actual test' is performed by operating the process to the trip point, and ensuring that the intermediate relays and instruments properly respond. The 'simulated test' is used in circumstances in which the instruments are of a nature that create a forced shutdown or start-up in a very hazardous process; or the instrument schemes are so complicated that the interlocks cannot be checked on a single-trip shutdown (Barclay, 1988).

The technical paper described the interlock testing programme as consisting of more than merely verifying that the interlock works. There is also inspection of the interlock equipment for deterioration. The 'function test' on instrumentation is required following mechanical work on one or more of the safety system elements. In short, all devices or systems that have been disturbed enough to affect their normal function will be function tested prior to being restored to service (Barclay, 1988).

Another company – a different emphasis on safety critical instrument systems

Gregory McMillan, who was an engineering fellow in process control at the Monsanto Chemical Company (US), and later a consultant, approached the subject differently. He made a very entertaining presentation entitled, 'Can you say "process interlocks"?' at the 1988 AIChE Loss Prevention Symposium (McMillan, 1987). It was a written as a parody of a popular youngsters' TV show.

Monsanto felt there were some real benefits in separating the economic issues from the safety issues and focusing on the interlocks that really require special attention. Prior to this separation, human life was grouped with major property protection and formed a large class for which it was difficult to enforce stringent design and test requirements (McMillan, 1987).

In earlier times, costs associated with business interruption which can exceed property losses after an interlock failure were largely ignored and those interlocks were often placed with the so-called 'operational interlocks'. McMillan proposed a system which classified the protection instrument loops into four classes:

100	Class I	Community protection (safety) loops
1000	Class II	Employee protection (safety) loops
5000	Class III	Major property and production (economic) loops
100 000	Class IV	Minor property and production (economic) loops

The article indicates that alarms and interlocks can proliferate to the point where they detract from the most critical loops. To be effective, a logical cause-and-effect approach to interlock classification is required. It must be determined which events are the direct and distinct causes of hazardous events.

In the script of this satirical article, Fellow, the process control engineer, explains that in the past, interlocks would have been placed on many but not all of the indirect causes of a release. In his example, a steam-heated chlorine vaporizer would have only two direct chemical process causes of a release. These direct causes are high pressure which can open a relief device or high temperature. The high temperature can accelerate corrosion.

In the past, this company's process control design would have placed interlocks on many of the indirect causes of releases. The indirect causes include a wide-open steam control valve, a closed chlorine gas valve downstream of the vaporizer, or a wide open upstream nitrogen regulator. Unfortunately, with the large number of interlocks with similar test requirements, there was not sufficient time and money to ensure the integrity of all of the interlocks.

McMillan suggests that the most direct and distinct cause of a potential release from this steam-heated chlorine vaporizer is high pressure, and high temperature is the second most likely cause. Thus, the high-pressure and high-temperature interlocks should receive the most severe classification and the most testing attention.

McMillan points to statistics that 80% of all interlock failures are due to failures of the field device. Those statistics show that 45% of the interlock failures are measurement device failures and the other 35% are caused by valve or valve actuator failures (McMillan, 1987).

Another approach to proof-testing in Louisiana

At PPG Industries, Lake Charles, Louisiana (US), there are numerous instrument loops which provide critical safety, alarm and shutdown functions. These protective instruments are located on reactors, oil heaters, incinerators, cracking furnaces, compressors, steam-heated vaporizers, kettles, distillation columns, and boilers. There are also process analysers and flammable vapour detectors providing process safety (Sanders and Wood, 1991).

McGraw-Hill Publications has granted special permission to use excerpts from a previous PPG Industries article in *Chemical Engineering* entitled,

'Don't leave plant safety to chance', Volume 98, No. 2, February 1991 (Sanders and Wood, 1991). Most of the material used in the rest of this chapter will be taken directly from the article and not be shown in quotes.

Instrument loops serving equipment described above can function in either an on-line or a standby manner; both types can fail. Failure of an on-line loop, such as a failure of a level control valve, becomes known rather quickly when the operation deviates either gradually or drastically from the normal. Depending on the type of failure, this may place a demand on the standby loop.

However, failure of a standby instrument loop, such as an alarm or safety interlock, will not become evident until a demand is placed on it to function. Problems developing in these loops must be discovered by periodic proof-testing.

Figure 8.6 Proof-testing a flammable gas detector in a tank farm

Figure 8.7 Proof-testing a high-pressure switch on a refrigeration unit

A proof-test programme cannot be left to someone's memory. It must follow a well-structured format to accomplish the essential steps, regardless of the myriad of other activities and distractions that tend to absorb all the supervisors' and mechanics' time. (Sanders and Wood, 1991).

What instruments are considered critical?

When PPG Lake Charles first initiated its proof-test programme, efforts to classify which safety devices were truly 'critical' were not specific enough. Hence, the original programme allowed too many instruments into the test system, which created a top-heavy burden. To prevent this from happening, the following information should be developed for critical loops (Sanders and Wood, 1991):

1. A listing of critical alarms and safety interlocks essential to safe plant operations

2. Criteria by which the truly 'critical instrument' system can be characterized as such, and distinguished from other important (but not critical) systems

The test procedure for complex interlocks should be developed with engineering help, and, for consistency, should be documented in writing. Any proposed additions or deletions should be screened by a third party group or committee. An appropriate authorization path – through which changes in shutdown set points and instrument values can be made in a systematic way and on the record – should be in place.

Test results must be recorded systematically, and must be easily accessible to all who need them for analysis. Defective or worn components should be identified and repaired immediately.

Management must take an active role in the stewardship of this programme. High levels of compliance with stringent test schedules should be urged and rewarded.

It takes years to develop and fine-tune a proof-test programme. Time is required to identify all of the instrument loops that need to be included, and more time is needed to systematically collect or develop data sheets containing basic operational information for each device or loop.

During the development stage, test methods need to be defined, and the appropriate personnel responsible for testing must be identified and trained. Finally, test frequencies must be decided.

The Lake Charles facility has over 2600 loops in its proof-test programme. In many cases, the test frequency is determined by decision-tree engineering studies. Critical instrument loops appear on a proof-test schedule, which is distributed monthly to the operations and maintenance departments (Sanders and Wood, 1991).

Prioritizing critical loops

Some process safety instrument loops are more critical than others, hence PPG Lake Charles, LA assigned priorities.

Priority 1
These are critical instruments whose failure would either cause, or fail to inform of, situations resulting in accidental fire, explosion, uncontrolled release of dangerous materials, reportable environmental releases, or major property or production loss. The alarms assigned a Priority 1 include those that have been mandated as such by: outside agencies; an in-house technical safety review committee; reliability studies; and specific alarms deemed critical by operations supervisors. All of these alarms are on a regular proof-testing schedule (Sanders and Wood, 1991).

The Lake Charles facility has about 1300 Priority 1 safety instrumentation loops. These include alarms and trips on storage tanks containing flammable or toxic liquids, devices to control high temperature and high pressure on exothermic reaction vessels, and control mechanisms for low-flow, high-temperature fluids on fired heaters. Other Priority 1 instruments include alarms that warn of flame failure on fired heaters, and vapour detectors for emergency valve isolation and sprinkler system activation.

Figure 8.8 A chemical process operator relies on a furnace fuel shutdown system

Figure 8.9 A chemical process operator witnesses the testing of an emergency isolation valve below a sphere

Priority 2
These are critical instruments whose failure could cause, or fail to inform of, serious conditions involving environmental releases, property or production losses, or other non-life-threatening situations. These alarms are given a slightly lower priority, but are also proof-tested on a regular schedule.

There are over 1300 Priority 2 alarms at PPG Lake Charles. Examples include those alarms or trips on refrigeration compressors, rectifiers, cooling towers, kettles and stills, and those controlling power and instrument air.

Priority 3
All other alarms which assist operations but are *not* considered critical devices are not on a regular proof-testing schedule.

Proof-test frequencies

Assigning proof-test frequencies for complex safety instrumentation loops requires 'sound engineering judgement' for simple systems. For more complex, interlock systems, the frequency is a function of the tolerable hazard rates. For example, DuPont Co.'s Sabine River Works (Orange, Texas, US) has 35 000 instruments in service. Every safety interlock is tested on a periodic basis, usually every 12 months, following a detailed written test procedure (Barclay, 1988).

Once a test frequency is established by DuPont for a particular interlock, a significant review and multiple approvals are required to permanently remove the interlock or change its test frequency.

Many of PPG's high-pressure and high-temperature alarms are tested every 6 months. About half of the PPG Lake Charles complex test frequencies have been developed using detailed reliability studies that consider the 'hazard rate' (the acceptable probability of a process accident) and the 'demand rate' (the number of times the critical alarm or shutdown function is required in service) (Sanders and Wood, 1991).

Within PPG's chlorinated hydrocarbon complex – which comprises a major portion of the Lake Charles facility, and produces eight different product lines – 26% of the loops are on a 1 year test frequency, 42% are on a 6 month frequency, 15% are tested every 4 months, 11% every 3 months and the remaining 6% at various other frequencies from weekly to every 2 years.

Administering the critical instrument proof-test programme

PPG Lake Charles' schedules for testing complex safety loops are distributed monthly in a packet to area supervisors. The packet includes a form listing the loops due in the current month (as well as those past due), the test frequency, and the authority requesting the test (Sanders and Wood, 1991).

Six independent groups of technical personnel provide proof-testing services. Proof-testing is performed by electricians, critical metering mechanics, refrigeration mechanics, analyser repair technicians, instrument maintenance personnel, and area instrument mechanics.

Figure 8.10 An instrument repairman proof-tests a critical intstrument panel

Once a loop has been proof-tested, the results are entered into the mainframe computer, along with the date tested and the condition found. All employees have 'read' access to this data on many terminals throughout the complex, but only authorized personnel have access to update the actual database.

Keeping track of other aspects of the system is also done on computer. For example, if a set point on a particular instrument needs changing, the request and approval (both approved by the area supervisor) are done at the computer terminal. Loops that are no longer in service can be deleted by the authorized personnel. New loops can be added by an individual designated by the technical safety committee, who reviews all new and modified installations.

Finally, monthly compliance reports are issued to top plant management. Particularly critical to the success of the programme is management's full support and commitment to compliance with process safety administration.

Additional information on mechanical integrity

The Center for Chemical Process Safety has recently released a new 'how to . . .' book (Center for Chemical Process Safety, 1992) which contains practical appendices on inspection and test procedures. These appendices are practices at major chemical manufacturing facilities. Chapter 8 should be consulted for specific information and actual plant programmes.

Appendix 8-B is entitled 'Example of Test and Inspection Equipment and Procedures'. This Appendix covers minimum testing and inspection requirements of all classified equipment. 'Classified' equipment is equipment identified in operational safety standards and governmental regulations. Appendix 8-C is called 'Example of Field Inspection and Testing of Process Safety Systems'. Appendix 8-E is named 'Example of Criteria for Test and Inspection of Safety' and in 17 pages it covers safety relief valves, rupture discs, control loop/manual actuated emergency vent devices and explosion vents.

These guidelines should be consulted if the reader is building or improving a mechanical integrity programme.

References

American Petroleum Institute (1989) *Pressure Vessel Inspection Code: Maintenance, Inspection, Rating, Repair, and Alteration*, API 510, 6th edn, API, Washington D.C.

American Petroleum Institute (1990) *Inspection of Piping, Tubing, Valves, and Fittings*, API 574, 1st edn, API, Washington D.C.

American Petroleum Institute (1991) *Tank Inspection, Repair, Alteration, and Reconstruction*, API 653, 1st edn, API, Washington D.C.

Barclay, D.A. (1988) Protecting process safety interlocks. *Chemical Engineering Progress* (February), 20–24

Braun, N. (1989) Silent sentinels: lessons on pressure relief valve performance. *Valve Magazine* (Spring), 28

Center for Chemical Process Safety (1992) *Plant Guidelines for Technical Management of Chemical Process Safety*, Center for Chemical Process Safety of the American Institute of Chemical Engineers, New York, Chapter 8

Institution of Chemical Engineers. *Safer Piping: Awareness Training for the Process Industries*, Hazard Workshop Module 12. Available as a training kit with a video, slides, a teaching guide, etc. from the Institution of Chemical Engineers, 165–171 Railway Terrace, Rugby CV21 3HQ, Warwickshire, UK

Krisher, A.S. (1987) Plant integrity programs. *Chemical Engineering Progress* (May), 21

McMillan, G.K. (1987) Can you say 'process interlocks'? *Technical Journal* (May), 35–44. Also presented at the AIChe Loss Prevention Symposium, New Orleans, March

Mueller, W.H., Arnold, C.G. and Ross, B.W. (1986) Reducing risks through a pressure vessel management program. AIChE Loss Prevention Symposium, New Orleans, April

National Association of Corrosion Engineers (1989) *A State-of-the-Art Report of Protective Coatings for Carbon Steel and Austenitic Stainless Steel Surfaces Under Thermal Insulation and Cementitious Fire Proofing*, NACE Publication 6H189, Technical Report by the National Association of Corrosion Engineers Task Groups 7-6H-31*

National Petroleum Refiners' Association (1990) *The Exchanger, Maintenance Information for the Hydrocarbon Processing Industry*, NPRA, Washington D.C.

Robertson, K. (1990) Responsible care is template for taking effective action. *CMA News*, **18**, No. 9, pp. 11, 12. Chemical Manufacturers' Association, Washington D.C.

Sanders, R.E. and Wood, J.H. (1991) Don't leave plant safety to chance. *Chemical Engineering* (February), 110–118

Sanders, R.E. and Woolfolk, W.H. (1984) Process safety relief valve testing. *Chemical Engineering Progress* (March), 60–64

Sanders, R.E. and Woolfolk, W.H. (1987) Dynamic testing and maintenance of safety relief valves. *Chemical Engineering* (October), 119–124

Van Boskirk, B.A. (1986) Don't let your plant blow up – an overview of the elements of overpressure protection. Joint Meeting of the Southeast Texas Section of the American Institution of Chemical Engineers and the Sabine-Neches Section of the ACS, April

Chapter 9

Properly managing change within the chemical industry

Preliminary thoughts on plant modification control

There must be a formal method to deal with change in the chemical industry. The safety that was designed into the original process was often obtained after a multidisciplinary design team agonized over the optimum arrangement of process and layout. This process safety must not be jeopardized by poor-quality modification schemes.

No recipe or procedure can be devised to be universally acceptable. The exact approach used to scrutinize a proposed change must be site specific and developed for that location. There must be a *sustained* management commitment to the management of change programme since this may require a change in culture within many organizations.

Each chemical plant and refinery must adopt or develop a procedure tailored to fit the specific hazards, the available technical resources, the culture of the organization and any required governmental regulations. It must be practical and workable *without undue delays*. To ensure the procedure continues to be properly utilized, there must be a periodic audit.

Essential elements of an effective management of change policy would include a programme in which *all employees*:

1. Understand the definition of change and why it is necessary to examine proposed changes
2. Recognize changes as they are proposed and seek third-party review
3. Have access to a qualified resource person and an available committee that can assist in identifying all potentially negative consequences of a proposed change
4. Document the changes on drawings (process, electrical, instrumental, electrical area classification, underground, etc.), revise operational procedures, change instrument testing methods, revise training manuals, etc., if necessary
5. Ensure that all recommendations offered to enhance process safety are studied and implemented in a timely manner
6. Believe that the company's management firmly supports the programme

133

A backward glance at earlier chapters

For a chemical manufacturing facility to survive in a dynamic industry, it must be able to quickly adapt to changing conditions such as increasing production, reducing operating costs, improving employee safety, accommodating technical innovation, compensating for unavailable equipment and/or reducing pollution potentials. The chemical plant must also have a method to review temporary repairs, temporary connections or deviations from standard operations.

It is essential that chemical plant modifications are properly engineered and implemented to avoid actual and potential problems. Chapter 2 demonstrates that a hidden practical or technical flaw in a noble effort to correct a certain problem can blemish a designer's reputation. It can also be dangerous.

Undesirable side effects may also be the result of modifications made in preparation for maintenance and/or activities in the implementation of maintenance. Some of these situations, previously discussed in Chapters 3 and 4, require more effort than the isolated projects. A system to review work orders and the preparations for maintenance requires continued diligence. Well-publicized, up-to-date mechanical and instrument specifications and written repair procedures, which include lists of blinds will reduce many misunderstandings and incidents.

Undesirable side effects can be even more easily the result of one-minute modifications, as discussed in Chapter 5. To address the problems of one-minute modifications, chemical plant management must persistently encourage employee awareness and train their employees about the potential dangers that can be created by the quick, inexpensive substitutions. It is essential that well-maintained engineering and equipment specifications are readily available. Changes which might include improper substitutes, such as incompatible materials of construction or improper procedures, must be reviewed by a third party. But this is sometimes easier said than done in the sometimes hectic pace of maintaining maintenance and production schedules.

It is crucial that companies refrain from making their management of change procedures so restrictive or so bureaucratic that individuals try to circumvent the procedures. Overcomplicated paperwork schemes and procedures which are perceived as ritualistic delay tactics must be avoided. It has become apparent that some companies require awkward, time-consuming review processes during the day shift. It has been said in those companies that the changes occur at night.

'Organizations do not have good memories; only people have good memories' is a quote from Trevor Kletz. Experienced employees can move due to promotions, changes in the organizational structure, retirements, or acceptance of other employment opportunities. When these well-trained individuals (be they auxiliary chemical process operators, lead operators, foremen, design engineers, etc.) are gone, all of that acquired experience becomes no longer available. Any such move could result in a loss of process safety knowledge. This type of change is often overlooked.

The industry 'head-count' (or the total number of skilled employees) is sacred in certain organizations. There is persistent management trend to be 'leaner and meaner' than the competition. A measurable excess in human resources is a luxury that is rarely found in today's chemical industry in developed countries. Management must constantly be attuned to the problems of dilution of knowledge by movement of their human assets.

Gradual changes created by unauthorized alterations, deterioration, and other symptoms of ageing can compromise the integrity of containment and protective systems. The presence of these unwanted 'modifications' can be minimized by proactive inspection, safety instrument system testing and the follow-up repairs. Chapter 8 covered these concepts.

Some of the 1970s chemical plants' approaches to plant changes

Four chemical corporations shared their progressive modification procedures in April, 1976, at a Loss Prevention Symposium sponsored by the American Institute of Chemical Engineers. Those technical papers were published in the AIChE's *Loss Prevention*, Volume 10.

Peter Heron described the approach that BP Chemicals International, Ltd, London used at that time. He pointed out that all process and plant changes were subjected to a regime of formal scrutiny and authorization. Minor alterations such as the addition of a valve, a change in materials of construction, or a switch in the type of mechanical seal were made at the discretion of the group manager, or the supervisory level above the unit supervisor. This individual had to satisfy himself that adequate consideration was given so that there were not any undesired side effects (Heron, 1976).

Major changes to existing units in BP Chemical, at the time the article was published, required consideration and formal authorization by personnel at the departmental management level. A procedure that was used at one of the BP chemical Plants was included in the article, and it included important items to be considered. Parts of two items on the list of 14 items are quoted here (Heron, 1976):

'. . . The proposal initiator and plant manager, plant engineer, chemical engineer and instrument/electrical engineer, as appropriate, should together discuss any proposed modification. They will prepare a set of notes and a sketch describing the modification and submit these for approval to the relevant staff.'

'. . . The plant manager will assess the effect of the proposals on all plant operations, including normal and routine operations, start-up, shutdown and emergency actions. He must also check that hazardous conditions cannot arise due to mal-operation.

The plant engineer and/or instrument electrical engineer will assess the effect of the modification on maintaining plant and equipment and will also ensure that the proposal meets:

a. The original plant design standards were appropriate
b. The level of good engineering standards demanded on site. . . .'

Peter Heron's article concludes with a statement that all of these procedures bring together a multidisciplined team, which ensures fewer problems in implementing, commissioning and operating modified units. It further reflects that there is often a dilution of experience in management and these procedures assist in ensuring consideration of a broad spectrum of potential side effects (Heron, 1976).

Trevor Kletz presented a number of very suitable examples of modifications that went sour in his article at that same symposium in 1976. After presenting several pages of problems with modifications, Kletz (1976) stated:

> 'None of the accidents described here occurred because knowledge was lacking on methods of prevention: they occurred because no one saw the hazards, nor asked the right questions. To prevent similar accidents in the future, a three-pronged approach is necessary.
>
> 1. There must be a rigid procedure for making sure that all modifications are authorized only by competent persons, who, before doing so, try to identify all possible consequences of the modification then specify the change in detail.
>
> 2. There must be some sort of guide sheet or check list to help people identify the consequences.
>
> 3. An instruction and aids are not enough. People will carry out the instructions and use the aids only if they are convinced that they are necessary. A training program is essential.'

Kletz's article also presented the 1976 procedures utilized by Imperial Chemicals Industries, Ltd (ICI) Wilton, UK. It stated that within the Petrochemicals Division of ICI, any modification, even if it is very inexpensive, or temporary, must be authorized in writing by a competent manager and an engineer (Kletz, 1976).

The ICI procedures required the use of a 'safety assessment' guide sheet, which discussed overpressure protection equipment, electrical area classification, changes in alarms and trips, and other categories that might diminish the safety of the system. The guide sheet also had questions on the relevant codes of practice and specifications, the materials of construction and fabrication standards, and any necessary changes in operating conditions. This ICI 'safety assessment' guide sheet, widely published first in 1976, seems to have withstood the test of time. It was published again in 1992 in a 'How to...' section of a current book of practical guidelines.

It would be unreasonable to think that both of these companies are still operating exactly as they did in the 1970s. Many chemical companies restructured and reduced the size of their technical staffs in the mid-1980s, and many other organizations added chemical process safety professionals. However, these approaches still seem very useful and can work.

How are chemical plants addressing plant modifications during the 1980s and beyond?

In 1985, the Canadian Chemical Producers Association (CCPA) released a pamphlet entitled, *Essential Components of Safety Assessment Systems*.

This pamphlet was developed to help Canadian chemical producers determine the adequacy of their process safety programmes. Modifications to a plant or process was one of the nine internal programmes examined by the CCPA. The guiding principles required a management programme to formally examine and approve any significant changes in chemical components, process facilities or process conditions whether temporary or permanent, prior to implementation. The procedure, as recommended by the CCPA, addressed 12 elements. It was intended that each element would be reviewed by qualified individuals to assess if the proposed change could jeopardize the integrity of the system. The following list is based upon the CCPA's publication and could be used as a simple evaluation form (Canadian Chemical Producers' Association, 1985).

1. Does the change involve any different chemicals which could react with other chemicals, including diluents, solvents and additives already in the process?
2. Does the new proposal encourage the production of undesirable byproducts either through the primary reactions, through side reactions or through introduction of impurities with the new chemical component?
3. Does the rate of heat generation and/or the reaction pressure increase as a result of the new scheme?
4. Does the proposed change encourage or require the operation of equipment outside of the approved operating or design limits of chemical processing equipment?
5. Does the proposal consider the compatibility of the new chemical component and its impurities with materials of construction?
6. Has the occupational health and environmental impact of the change been considered?
7. Has the design for modifying the process facilities or conditions been reviewed by a qualified individual using effective techniques for analysing process hazards, particularly when the modifications are being made in rush situations or emergency conditions?
8. Has there been an on-site inspection by qualified personnel to ensure that the new equipment is installed in accordance with specifications and drawings?
9. Have the operating instructions and engineering drawings been revised to take into account the modifications?
10. Have proper communications been made for the training of chemical process operator, maintenance craftsman, and supervisors who may be affected by the modification?
11. Have proper revisions been made to the process control logic, instrumentation set points and alarm points, especially for computer control systems, to properly respond to the modification?
12. Have provisions been made to remove or completely isolate obsolete facilities in order to eliminate the chances for operator errors involving abandoned equipment?

The CCPA's publication is an excellent 44-page booklet which lists 58 references covering emergency planning, process hazards reviews, fault tree analysis, evaluation of toxic vapour cloud hazards, etc.

The Center for Chemical Process Safety

The American Institute of Chemical Engineers (AIChE) has been involved in chemical process safety and loss prevention for chemical and petrochemical plants for a number of years. In early 1985, the AIChE established the Center for Chemical Process Safety (CCPS) to intensify development and dissemination of the latest scientific and engineering practices for prevention and mitigation of catastrophic incidents involving hazardous chemicals. The CCPS serves as a focus for developing literature and courses to continue to improve chemical process safety. In one of the CCPS's newer books, *Guidelines for Technical Management of Chemical Process Safety*, 1989, there was a review of technical safety management systems of selected large companies.

This survey indicated that only one major chemical company out of seven rated themselves with a fully implemented programme when it came to 'management of change' (Center for Chemical Process Safety, 1989). The other companies indicated that some elements of the 'management of change' programme have been implemented.

During a February 1992 CCPS-sponsored 'Process Safety Management Course', the 'management of change policy' was identified by a cross-section of industry, insurance and consulting firms as the most difficult to implement of all the elements of the new OSHA Process Safety Standard (1910.119). It was considered difficult because so many people were involved in implementation; accountability and responsibility were not defined; the extent of application was not defined; and it was contrary to many chemical plant cultures.

New recommendations and new regulations

In December 1988, the Organization Resources Counselors, Inc. (ORC), as a means to assist the US Occupational Safety and Health Administration (OSHA) prepared a report entitled *Recommendations for Process Hazards Management of Substances with Catastrophic Potential* (Organization Resources Counselors, 1988). This document was drafted to help OSHA revise standards for handling hazardous materials.

ORC is a Washington, DC private industry group with many representatives from the chemical industry. The ORC agreed to help because they were concerned about another worldwide disaster that might cause the US Congress to quickly develop legislation which could be ineffective. The ORC was also concerned about the proliferation of similar regulations at the state level and they felt a performance standard was essential if the law was to be effective (Boyen, 1992).

The ORC indicated that their report was capable of being a useful starting point in the development of a policy or programme for dealing with chemical process hazards. The report specifically indicated that it was intended as a guideline or starting point only, and it should be extensively reviewed and analysed before being implemented.

The American Petroleum Institute developed API 750 (American Petroleum Institute, 1990) and released it in January 1990. This practical 16-page document is based in part on the ORC report.

The Chlorine Institute (1990) developed and released Pamphlet 86, *Recommendations to Chlor-Alkali Manufacturing Facilities for the Prevention of Chlorine Releases* in October 1990. This 15-page pamphlet acknowledges that the ORC report *Recommendations for Process Hazards Management of Substances with Catastrophic Potential* was one of the primary documents used to develop it.

The ORC document emphasized the application of management control systems to facilities processing highly hazardous chemicals. The ORC report helped form a new law within the United States. The new standard, OSHA 1910.119, 'Process Safety Management of Highly Hazardous Chemicals,' is part of the US Code of Federal Regulations and was drafted in July 1990 and scheduled for implementation on 26 May 1992.

The section of the 'final rule' of the OSHA Process Safety Management law which addresses 'management of change' is found in paragraph (l) (US Department of Labor, 1992) and states:

'(1) The employer shall establish and implement written procedures to manage changes [except for "replacements in kind"] to process chemicals, technology, equipment and procedures; and changes to facilities that affect a covered process.

(2) The procedures shall assure that the following considerations are addressed prior to any change.
(i)   The technical basis for the proposed change;
(ii)   Impact of change on safety and health;
(iii) Modifications to operating procedures;
(iv)   Necessary time period for the change; and,
(v)   Authorization requirements for the proposed change.

(3) Employees involved in operating a process and maintenance and contract employees whose job tasks will be affected by a change in the process shall be informed of, and trained in, the change prior to start-up of the process or affected part of the process.

(4) If a change covered by this paragraph results in a change to the process safety information required by paragraph (d) of the section shall be updated accordingly.

(5) If a change covered by this paragraph results in a change to the operating procedures or practices required by paragraph (f) of this section, such procedures or practices shall be updated accordingly.

(Inclusion of the associated paragraphs (d) and (f) is beyond the intended scope of this book.) The 'final rule' also defines 'replacement in kind' as a replacement which satisfies the design specification.

Appendix C to OSHA 1910.119 is entitled 'Compliance Guidelines and Recommendations for Process Safety Management (Nonmandatory)'. It serves as a nonmandatory guideline to assist with complying to the standard. The appropriate section is quoted here (US Department of Labor, 1992):

' "Managing Change.' To properly manage change to process chemicals, technology, equipment and facilities, one must define what is meant by change. In this process safety management standard, change includes all modifications to equipment, procedures, raw materials and processing conditions other than 'replacement in kind.' These changes need to be properly managed by identifying and reviewing them prior to implementation of the change. For example, the operating procedures contain the operating parameters (pressure limits, temperature ranges, flow rates, etc.) and the importance of operating within these limits. While the operator must have the flexibility to maintain safe operation within the established parameters, any operation outside of these parameters requires review and approval by a written management of change procedure.

Management of change covers such as [*sic*] changes in process technology and changes to equipment and instrumentation. Changes in process technology can result from changes in production rates, raw materials, experimentation, equipment unavailability, new equipment, new product development, change in catalyst and changes in operating conditions to improve yield or quality. Equipment changes include among others change in materials of construction, equipment specifications, piping prearrangements [*sic*], experimental equipment, computer program revisions and changes in alarms and interlocks. Employers need to establish means and methods to detect both technical changes and mechanical changes.

Temporary changes have caused a number of catastrophes over the years, and employers need to establish ways to detect temporary changes as well as those that are permanent. It is important that a time limit for temporary changes be established and monitored since, without control these changes may tend to become permanent. Temporary changes are subject to the management of change provisions. In addition, the management of change procedures are used to insure that the equipment and procedures are returned to their original or designed conditions at the end of the temporary change. Proper documentation and review of these changes is invaluable in assuring that the safety and health considerations are being incorporated into the operating procedures and the process.

Employers may wish to develop a form or clearance sheet to facilitate the processing of changes through the management of change procedures. A typical change form may include a description and the purpose of the change, the technical basis for the change, safety and health considerations, documentation of changes for the operating procedures, maintenance procedures, inspection and testing, P&IDs, electrical classification, training and communications, pre-startup inspection, duration if a temporary change, approvals and authorization. Where the impact of change is minor and well understood, a check list reviewed by an authorized person with proper communication to others who are affected may be sufficient. However, for a more complex or significant design change, a hazard evaluation procedure with approvals by operations, maintenance, and safety departments may be appropriate. Changes in documents such as P&IDs, raw materials, operating procedures, mechanical integrity programs, electrical classifications, etc. need to be noted so that these revisions can be made permanent when the drawings and procedure manuals are updated. Copies of process changes need to be kept in an accessible location to ensure that design changes are available to operating personnel as well as to PHA team members when a PHA is being done or one is being updated.'

The AIChE's Center for Chemical Process Safety developed a 'how to . . .' type of book which addresses most of the concerns of the OSHA's proposed Process Safety Management law and all of the concerns of management of change. This book is useful to the unit foreman, area supervisor, superintendent and the manager of a facility that manufacturers, handles or stores hazardous chemicals. It is a must for the 'on-site' organization which is developing the specific procedures of a management of change programme.

Plant Guidelines for Technical Management of Chemical Process Safety (Center for Chemical Process Safety, 1992a) provides contributions from many participating chemical manufacturers. There are numerous check-lists, decision trees, proposed forms, etc., and nine very practical appendices in Chapter 7, 'management of change'.

An overview of training in a management of change programme

Since changes may be introduced into a process or facility, by supervisors, engineers, chemical process operators, process control technicians, electricians, machinists, pipefitters, other craftsmen, purchasing agents, contractors, etc., there must be a training programme to reach all of the players on the team. Employees must understand what is meant by change, and companies must convince all of the players concerned that these modification procedures are necessary. There must be a programme to identify changes and encourage the proper individuals to review such change, and there must be some type of assessment guide sheet or trained 'modifications man' to help the modification's sponsor to logically study the proposal.

One way to assist in properly training key individuals within a plant site is to purchase the Institution of Chemical Engineers *Hazards of Plant Modifications* (Kletz, 1976). The effectiveness of this training can be greatly enhanced if case histories of the accidents caused by an improper plant modification at your plant site are added.

Previous incidents in which a plant modification was suspected to be a contributing culprit should be researched for written reports, photos, sketches, etc., and this local incident included in the training. Perhaps some of the incidents addressed in this book can help make special points. Naturally, any new incident within your organization or reported by the news media can help training efforts.

A workable approach for reviewing proposed plant modifications

Note: As of this writing the OSHA 'Final Rule' of 'Process Safety Management of Highly Hazardous Chemicals . . .' has been only published for about a month. The approaches discussed within this chapter are based upon my 18 years of Loss Prevention experience in a major chemical plant. This chemical plant processes toxic gases and liquids, flammable gases, flashing flammable liquids and combustible liquids.

My experiences are from within a chemical complex that made detailed process hazards analysis 20 years before the law required such studies and has had a team of two or more seasoned process engineers performing quantitative process hazards analysis and updates for over 10 years. This plant had a huge maintenance budget of over a million dollars a week and I believe it is not desirable to document every trivial change. I firmly believe that OSHA does not want to direct the resources of a corporation into extensive record keeping on trivial changes which meet or exceed plant specifications, but it is difficult to describe a programme.

I have interpreted the law to encourage the following workable program. However, the author, editor and publisher specifically disclaim that heeding any of these methods will make any premises or operations safe or healthful or in compliance with the OSHA Law.

For a workable approach, three 'C's' are essential. These are Commitment, Culture and Cash.

There must be commitment and direction from management. Management at each level in the organization must visibly support and continuously reinforce the policies which are designed and implemented to reduce spills, releases, fires and explosions. There must also be clear roles and responsibilities within process safety.

A logical approach to plant modifications must be developed for each chemical plant site. Ideally, it should be a tiered approval system. Careful consideration must be given to the size of the facility, the relative hazards of the chemicals, the type of equipment and the number of employees, including process safety personnel, engineers, etc.

To change the culture of the organization, all plant employees which can impact on change must be trained to understand what is considered to be a plant modification. Each individual considering a small change should be encouraged to discuss this idea with a peer to question if the change is considered within specified operating limits or acceptable maintenance or engineering practices. This is not always practical.

Employees should understand the review process; see a tiered approach in Figure 9.1. The unit manager (area superintendent) should be first consulted. This may be triggered by discussions, a work order approval procedure which requests permission for 'changes' or other means. This approach assumes that the unit manager is a second level supervisor with one or more operations and/or maintenance foremen reporting to him.

The nonmandatory section of the 'Management of Change' states that organizations must define what is meant by change. The unit manager should decide if the change is within specifications, or within the boundaries of 'replacement in kind', basic good engineering judgement or normal variances allowed.

Some examples of 'replacement in kind', 'satisfies plant specifications' or basic good engineering judgements that can be approved by the unit manager include:

1. Replacement of gate valves with ball valves (within the plant valve specifications or within regular usage for that service)

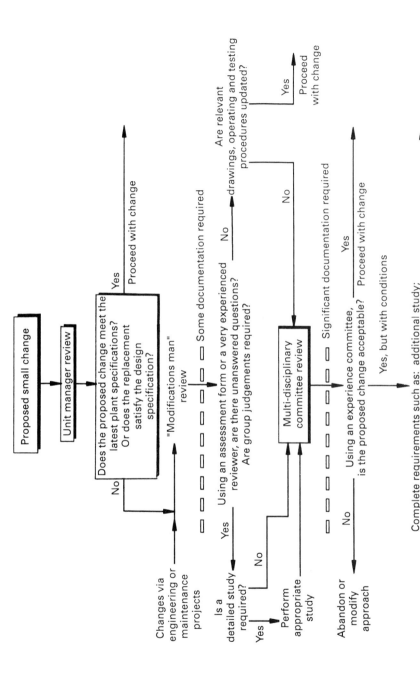

Figure 9.1 A management of change decision tree

2. Replacement of compressed asbestos gaskets with a different gasket material which has been approved by engineering and written into the plant specifications
3. Replacement of heat exchanger tubes with a more corrosion resistant material which has been approved by an experienced corrosion engineer
4. Changes in lubricants for pumps and valves which have been approved by an experienced maintenance engineer

Some changes that could be considered a significant modification are listed below:

1. Mechanical, process or instrument system changes made to increase production rates
2. Facility changes made to significantly increase storage capacity of a hazardous material
3. Feedstock, equipment or procedural changes made to increase production yield or increase product purity
4. Alterations to protective equipment systems, such as changes to critical values for alarm, interlock or shutdown systems; or changes involving safety relief or vent systems
5. Changes made to compensate for unavailable equipment, instruments, rotating equipment or vessels, such as using jumpovers, hoses, vacuum trucks, etc.
6. Experiments to change raw materials, additives, inhibitors, catalyst, etc.
7. Proposals to operate at significantly different pressures, temperatures, flow rates, acidity, etc., which are considered outside the well understood and documented 'safe operating envelope'
8. Proposals to improve personal safety or process safety, or for increased environmental stewardship
9. Plans to change materials of construction for major components and lesser components such as new gasket materials which are under experimental service conditions
10. Unique 'one-time maintenance' such as chemical cleaning, hot tapping, freezing pipelines to work downstream of the plug
11. Schemes for coping with temporary situations such as pipe clamps on leaking high hazard lines, operating with a heat exchanger out of service, changing delivery methods, e.g. accepting a truck delivery when the normal method was by drum or pipeline
12. Restart of a unit after being idle for 6 months or more
13. Decommissioning and demolition of parts of units
14. Any change in the physical plant which could increase business interruption potential which may not have been identified above
15. Any item that the unit supervisor or the process safety engineer believes requires additional examination

The third 'C' (cash) or sufficient resources must be allocated to provide further review if necessary. A 'modifications person' should be available to operations, engineering and maintenance. This modifications person may be a process safety engineer, loss prevention engineer or a mechanical or chemical engineer who has been trained in chemical process safety. Preferably, this individual would be a plant employee, but he could be a company regional engineer, a property insurance consultant, or a contractor. The modifications person must understand the basic loss prevention principles of proper layout, fundamentals of fire and explosion protection, overpressure protection, electrical area classification, property insurance guiding principles, and the like. It is unrealistic to have such a well-trained individual who can think of all the right questions, so 'A safety assessment checklist for modifications' should be utilized. Table 9.1 was adapted from Kletz (1976), other published sources and through the courtesy of unnamed progressive chemical manufacturers and petroleum refiners.

Table 9.1 A safety assessment checklist for modifications

Change Register No.:............................. Date Submitted ....
Title of Proposal: ....
....
Specific Plant Area ....
Proposed Start-Up Time: .... Rush?.......................
Description of the Proposal: ....
....
....
....
Reason for the Change: ....
....
....
....

Circle those factors which may be changed by the proposal

Proposed Classification
Capital Improvement
Environmental Improvement
Process Change
Abnormal Operations
Emergency Operations
Temporary Change
Materials Change
Preparation for Maintenance

Maintenance Considerations
Equipment Inspection
   Pre-modification
   Periodically in Service
Trip and Alarm Testing
Maintenance Procedures
Specialty Contractors
   Hot Tapping or Stopple
   On-stream Leak Repair
   Line Freezing
   Vessel Alteration

Process Conditions
Temperature
Pressure
Vacuum
Flow
Level
Composition
Flash Point
Reactive Conditions
Toxicity
Corrosion Potentials

Engineering Considerations
Instrument Drawings
Process Drawings
Wiring Diagrams
Trip & Alarm Procedures
Plant Layout
Pressure Relief Design
Flare & Vent Specifications
Design Temperature
Isolation for Maintenance
Static Electricity
Drainage

Table 9.1 cont.

Does the proposal properly address these *process concerns*:

1. Does the proposal introduce new chemicals in the form of new reactants, solvents, catalysts, or impurities?
2. If so, are the new chemicals flammable, explosive, toxic, carcinogenic, irritants, capable of decomposition, oxidants, etc.? If so, are safety data sheets available?
3. Does the rate of heat generation and/or reaction pressure increase as a result of this new scheme? Is there a potential for overtemperature during start-up, shutdown, normal operation or other cases such as loss of agitation, loss of utilities?
4. Are the vent and pressure relief systems sufficient under the new conditions?
5. Is there a risk of creating a damaging vacuum condition?
6. Is there an increased risk of backflow or cross-contamination?
7. Does the proposal introduce flammable liquids or gases or combustible dusts into areas which do not have the proper electrical area classifications?

Does the proposal properly address the *equipment/hardware concerns*:

1. Does the change involve the alteration of a pressure vessel? And if so, is the code certification preserved?
2. Is there sufficient pressure difference between the new operating pressure and the maximum allowable working pressure of the vessel?
3. Is the relief capacity adequate for process upsets, valve or tube failure, fire, loss of utilities, etc.?
4. Are remote-operated isolation valves now needed? Are 'double block and bleeds' required?
5. Have safety critical process alarms and shutdown systems been modified to include the new situation?
6. Does the proposal introduce a source of ignition? (including hot surfaces, flame, mechanical sparks, static electricity, electrical arcing, etc.)
7. Will the gas detection systems, fire-water systems, diking or drainage need to be changed to accommodate the change?

Does the proposal properly address the *procedural, training and documentation requirements*?

1. Have the process, mechanical and instrument drawings been updated, where required?
2. Have the new Material Safety Data Sheets been provided to Operations and Maintenance?
3. Have the start-up, normal shutdown and emergency shutdown scenarios and procedures been reviewed?
4. Have the schematic wiring and other electrical drawings been updated?
5. Have the equipment files been updated to show the addition of pressure vessels, storage tanks or revisions to them?
6. Have the sewer and underground drawings been updated, where required?
7. Have the alarm listings and safety critical proof-test procedures been developed?
8. Have all the other necessary maintenance testing and inspection procedures been developed?

Table 9.1 gives a few general questions. A chemical plant trying to develop a set of questions for a specific location should consider the 45 pages of excellent questions in the *Guidelines for Hazard Evaluation Procedures* (Center for Chemical Process Safety, 1992b).

The modifications person should be asked to review all modifications that unit supervisors feel need additional scrutiny. If the type of modification under consideration is not covered by plant specifications, codes or design and operating philosophy, or if unanswered questions are generated by an assessment form, then it is a candidate for further review and further approval levels.

Eight other examples of 'Management of Change Policies', 'Safety Assessment Checklists', 'Control of Change – Safety Management Practices', 'Change of Process Technology', etc. are found in *Plant Guidelines for Technical Management of Chemical Process Safety* (Center for Chemical Process Safety, 1992a).

If additional examination seems required then this review can be achieved most effectively with a hazards identification procedure or a multidisciplinary committee. Most medium-sized and large chemical and petrochemical corporations have or should implement flexible procedures for several layers of process safety reviews for capital projects such as major modifications, expansions, etc. There should be a procedure to alert or assign a modifications person to examine the early stages of the proposed change. If the change is covered by specifications or plant policy, or properly addressed by codes of practice, the review may stop at this point with or without a brief note, depending upon systems.

An abundance of ritualistic paperwork or overcomplicated paperwork schemes must be avoided. One experienced project engineer stated, 'It is easier to get forgiveness than permission.' Do not create or nurture such a system. Some published articles covering management of change systems and some actual plant procedures which were reviewed appeared too cumbersome to be effective.

If a proposed change falls outside of plant specifications, codes of practice, etc., then a chemical process safety review team evaluation may be in order. In the early 1990s many of these process safety review teams which are assigned the responsibility to address change carry names such as Safe Operations Committee, Hazards Review Committee, Technical Support Team, Facility and Operations Change Review Committee, Chemical Process Safety Review Team, etc. Typically, this technical safety committee is chaired by mid-level supervision such as a technical manager, engineering superintendent, manager of process safety, etc., who are not directly affected by the budgetary constraints or the start-up deadlines that the typical production department faces.

Some successful chemical process safety review committees have a nucleus of two or three individuals supporting the committee chairman. A process engineer from engineering, an operations representative and either a process safety coordinator, a loss prevention engineer or a well-rounded process engineer are often key central players in such a committee.

The committee should be rounded out with more individuals who can make specific contributions such as, but not limited to, process control engineers, mechanical engineers, unit supervisors, chemical process operators, chemists, environmental engineers, project engineers, fire protection engineers, reliability engineers, and personal safety personnel. Progressive organizations include the chemical process operators, shipping loaders, and other 'hands-on' personnel if the occasion warrants their presence.

In certain cases, such as changes created by a significant expansion, it is better to have a small group of specialists first identify the potential hazards and sometimes quantify the risk, prior to any type of committee review. Brief descriptions of various approaches to hazard identification are presented within the next section.

The proceedings of chemical process safety committee meetings should be captured in minutes. In many organizations there should be a review team of senior management who review the minutes and approve the conclusions. Once the minutes are accepted, the commitments made by various participants should be recorded and the progress of each recommendation should be periodically reviewed until such items are accomplished.

How should the potential hazards be identified and evaluated?

Frank Lees, starts off Chapter 8 'of *Loss Prevention in the Process Industries* (Lees, 1980) with these three paragraphs:

'The identification of areas of vulnerability and of specific hazards is of fundamental importance in loss prevention. Once these have been identified, the battle is more than half won.

Such identification is not a simple matter, however. In many ways it has become more difficult as the depth of technology has increased. Loss prevention tends increasingly to depend on the management system and it is not always easy to discover the weaknesses in this. The physical hazards also no longer lie on the surface, accessible to simple visual inspection.

On the other hand, there is now available a whole battery of safety audit and hazard identification methods to solve these problems.'

No single identification procedure can be considered the 'best' for all companies or all situations. Large refineries and large single train industrial chemical producers with limited products and large technical staffs will by nature approach their reviews differently from specialty batch operations for making limited campaigns of products which are specialty solvents, herbicides, insecticides, etc.

Some considerations for deciding on the type of evaluation would include:

1. Potential 'worst case' consequences
2. Complexity of the process or facility
3. Experience level of the available review members
4. Time and cost requirements
5. Corporate standards, state and local requirements

Any and all reviews should take an orderly, systematic approach. A general qualitative review may be sufficient, except for some sophisticated portions of the proposed modification. The review team should be responsible for determining deficiencies, but not defining solutions. Lees (1980) notes many selected references to help evaluate processes.

The AIChE's *Guidelines for Hazard Evaluation Procedures* (Center for Chemical Process Safety, 1992b) offers a wide variety of alternatives to review systems for hazards. These review procedures can be used to evaluate some plant modifications. Two basic categories of evaluations are: (1) adherence to good engineering practice and (2) predictive hazard evaluation.

Adherence to good engineering practice

Adherence to good engineering practice means that the modification is reviewed and compared to corporate or plant standards, state and local codes, insurance regulations and societal codes. Many companies have developed excellent checklists or standards based upon years of experience with the manufacture, use and handling of various chemicals.

Checklists, 'what if' methods, and multiple discipline reviews are often an excellent method to review a modification. These methods can ensure that there is adherence to design specifications, that previously recognized hazards are identified and that P&IDs and operating procedures are updated.

'What if' type studies have been used to some degree for years. This type of questioning activity was often the method a junior process engineer would experience when presenting a proposal to a review team in previous years. Not much has been written on this type of group brain-storming activity; however, (Center for Chemical Process Safety, 1992b) offers examples of a systematic approach. The method is suitable for an experienced staff reviewing a modification.

Checklists can be user friendly if properly prepared by experienced engineers. Such checklists can be very useful tools to assist less experienced engineers in considering situations that fault tree analysis might find if given enough time or that 'what if's' might overlook.

A brief one page memory jogging safety assessment checklist was made available by Trevor Kletz (Kletz, 1976). After a review of several management of change policies from several major companies, it appears that Kletz's checklist or a similar checklist was used as a basis for a few companies' procedures.

Lees (1980) has catalogued and presented a number of checklists. Lees (1980) states:

'One of the most useful tools of hazard identification is the checklist. Like a standard or code of practice a checklist is a means of passing on hard-won experience. It is impossible to envisage high standards in hazard control unless this experience is effectively utilized. The checklist is one of the main tools available to assist in this.'

'. . . Checklists are only effective if they are used. There is often a tendency for them to be left to gather dust on the shelves. This is perhaps part of the reason for the development of other techniques such as hazard and operability studies . . .'

Center for Chemical Process Safety (1992b) also offers 45 pages of sample questions in Appendix B. These questions cover process, (i.e. flowsheets and layout), equipment (reactors, heat exchangers, piping, instrumentation, etc.), operations, maintenance, personnel safety and other broad areas.

Other practical engineering information which can help evaluate modifications may be obtained from individual organizations such as the American Institute of Chemical Engineers, the American Petroleum Institute, the Institution of Chemical Engineers (UK), the National Fire Protection Association, etc. Chapter 10 lists a number of such resources and the postal addresses to obtain helpful information.

Predictive hazard evaluation procedures

Predictive hazard evaluation procedures may be required when new and different processes, designs, equipment, or procedures are being contemplated. The Dow Fire and Explosion Index provides a direct method to estimate the risks in a chemical process based upon flammability and reactivity characteristics of the chemicals, general process hazards (such as exothermic reactions, indoor storage of flammable liquids, etc.) and special hazards (such as operation above the flash point, operation above the autoignition point, quantity of flammable liquid, etc.). Proper description of this index is best found in the 57-page *Dow's Fire and Explosion Index, Hazard Classification Guide*, 5th edn, AIChE, New York, 1981.

The HazOp Study is a very popular predictive method which was developed in the Mond Division of ICI during the 1960s. A HazOp (Hazard and Operability) study is an analysis method for identifying hazards and problems which prevent efficient operation. Trevor Kletz was an early promoter of the HazOp method and in one of his recent books (Kletz, 1990) he states:

'A Hazard and Operability study (HAZOP) is the preferred method, in the process industries, of identifying hazards on new or existing plants.

'. . . We need to identify hazards before accidents occur. *Check lists* have the disadvantage that new hazards, not on the list, may be overlooked so we prefer the more open-ended hazop technique. It allows a team of people, familiar with the design, to let their minds go free and think of all the deviations that might occur but it is done in a systematic way in order to reduce the chance of missing something.

Note that hazop identifies operating problems as well as hazards.

Although hazop has been used mainly in the oil and chemical industries it can be applied to many other operations.

The technique is applied to a line diagram, line-by-line. Using the guide word 'NONE' we ask if there could be no flow, or reverse flow, in the first line. If so, we ask if this would be hazardous or would prevent efficient operation. If it would, we ask what change in design or method of operation will prevent no flow, or reverse flow (or protect against the consequences). Using the guides, 'MORE OF' and 'LESS OF' we ask if there could be more (or less) flow, pressure or temperature in the line and if this would be hazardous, etc. Using the guide words 'PART OF' and 'MORE THAN' we ask about the effects of changes in concentration or the presence of additional substances or phases. The guideword 'OTHER' reminds us to apply our questioning to all states of operation, including start-up, shutdown, catalyst regeneration and so on and we also ask if the equipment can be safely prepared for maintenance. We then study the next line in the same way. All lines should be studied including *service lines* and *drains*.'

A HazOp study team is best when it is a multidisciplinary group formed to bring a broader base of experience to the review. A HazOp may not be suitable for the small modification of a few extra valves or a new run of pipe, but it is desirable for the medium-sized and larger modifications.

Johnson and Leverenz (1992) stated that 'The hazard and operability studies (HAZOPS) method has likely become, over a period of less than ten years, the most widely used hazard evaluation procedure in the process industries.' They also explained that HazOps is a relative latecomer to the United States and it has attained a high degree of prominence in the US process industries. The method began in the UK and has now spread throughout western Europe and North America (Figure 9.2).

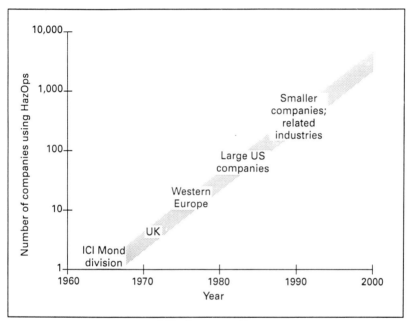

Figure 9.2 Estimate of extent of HazOps usage (Courtesy of R. W. Johnson, Battelle Memorial Institute)

Another widely-used predictive method is the use of fault tree analysis. This is a 'reverse-thinking' method. The analyst assumes an accident or specific undesirable event, the so-called 'TOP event'. This could be the release of a toxic gas from a reactor safety relief valve.

The analyst must then define the various equipment failures and human failures that could lead up to that event. For the release of toxic gas from a reactor safety relief valve, the analyst may consider loss of cooling and the operator ignoring the high-pressure alarm. Another path to this leak of toxics may be a double charge of one of the energetic reactants.

Failure rates for both equipment and people responses are assigned and the frequency and severity of a TOP event can be calculated. Should the risk be found to be unacceptable, additional process safety hardware or additional procedures can be recommended. Then, calculations can be made to determine the benefits of the additional hardware or procedures. The fault tree analysis method of evaluation is very sophisticated and a detailed explanation is beyond the scope of this book.

Variances, exceptions, and special cases of change

All deviations from normal plant policy requirements must follow a prescribed procedure. They must be approved by authorized individuals and properly documented prior to implementation. In ideal situations, all variances are submitted in writing.

A procedure to address taking alarms, instruments, or shutdown systems out of service

This standard 'safe operating procedure' was developed to create a uniform method to ensure that appropriate steps are taken prior to bypassing or removing an alarm, instrument or shutdown system from service. The procedure was developed to provide an effective way of communicating the status of an impaired instrument. The procedure has been in use for more than 3 years. It assumes that all instrumentation has been classified into three priorities.

The three priorities of safety critical systems at this plant were classified as (Sanders and Wood, 1991):

Priority 1 Safety critical instruments whose failure *would* either cause, or fail to inform of, situations resulting in accidental fire, explosion, uncontrolled release of dangerous materials, reportable environmental releases, or major property or production losses. The safety critical instruments assigned a 'Priority 1' rating include those that have been mandated as such by: regulating agencies, an in-house technical safety review committee and reliability studies, and include specific shutdown systems and specific alarms deemed critical by operations supervisors.

Priority 1 safety instrumentation loops include alarms and trips on storage tanks containing flammable or toxic liquids, devices to control high temperature and high pressure on exothermic reaction vessels, and control mechanisms for low-flow, high-temperature fluids on fired heaters. Other Priority 1 instruments include alarms that warn of flame failure on fired heaters, and vapour detectors for emergency valve isolation and sprinkler system activation. All of these alarms, shutdown valves and other critical instruments are regularly proof-tested on a well-defined schedule.

Priority 2 Safety critical instruments whose failure *could* either cause, or fail to inform of, serious conditions involving environmental releases, property or production losses, or other non-life-threatening situations. These instruments are given a slightly lower priority, but are also proof-tested on a regular schedule.

These Priority 2 safety critical instruments include alarms or trips on refrigeration systems, rectifiers, cooling towers, kettles, and stills (Kletz, 1990).

Priority 3 Instrument systems that are used to alert the chemical process operator of a non-hazardous abnormal condition that might otherwise be undetected. The failure to react to one of these alarms may create an off-specification product such as a low temperature alarm on certain distillation columns. These systems are not included in the proof-test programme.

1. *If an instrument or instrument system malfunctions, the operator tries to correct the problem.* If an alarm, instrument or shutdown system malfunctions, the operator's first response should be an attempt to restore it to service. It may be a plugged impulse line, flow shut off to an analyser, etc.

If the instrument system is a priority 1 or 2, go to step 3. If the alarm is a priority 3, go to step 2.

2. *The lead operator should be consulted and follow established procedures and his knowledge of the unit in each situation and determine the temporary steps required.* At the minimum, a note in the maintenance log book must be made to initiate repairs.

Stop here for priority 3 alarms.

3. *If the instrument system is considered a priority 1 or priority 2, lead operator must be consulted about the situation.*
4. *The lead operator will decide if it is safe to take the instrument out of service. If it is safe to do so, the lead operator will complete an 'out-of-service' tag and mount it on the control panel.* This is a small brightly coloured tag with adhesive on the back, like a 'Post It' tag. It should be mounted on the control panel or other prominent place.

   At the minimum, the lead operator will make a note in the operations log book and the maintenance log book. An orderly shutdown may be required; if so the shift supervisor should be consulted.

   If the situation arises which requires an instrument to be out of service for a very short period of time, such as an instrument mechanic freeing a stuck shutdown valve with the operator present, or temporarily pulling alarm cards to help identify 'mystery alarms', the lead operator should be informed and the operator should give his undivided attention to the problem, but an 'out-of-service' tag may not be filled out.

   The 'out-of-service' tag is not a substitute for promptly performing maintenance. The tag serves only to effectively communicate the status of instruments in need of repair.
5. *The lead operator must take immediate action to restore the out-of-service instrument.* On a normal day shift, this will require contact with the supervisor and the supervisor may direct the efforts of a mechanic, or generate an 'emergency' work order. On the back shifts or holidays, this may mean having the shift supervisor call out the mechanic(s) from home.
6. *When the repairs are successfully completed and the safety critical instrument is restored, the lead operator will remove the 'out-of-service' tag from the control board.* He fills out the blank which requests the date when the instrument was restored and tapes the small slip into the log book.
7. *A 'weekly out-of-service tagging checklist' will be completed by each unit every Monday morning before the day shift arrives.* This is a check of all

the panel alarm lights and a review of any work that remains to be accomplished.

As a cross-check, a review of all the 'out-of-service' tags of the previous week will be made. Each of the safety shutdown systems valves that has been bypassed during the previous week will be inspected in the field to ensure that it is neither blocked nor operating with the bypass open.

Safety critical instrument setting changes

Critical values of operating temperatures, pressures and flows are often established prior to the start-up of the unit. As experience is gained, it may be necessary to fine-tune the alarm points and shutdown values. At a minimum, this should be approved by the unit manager or area supervisor.

If the shutdown system is complex, or there are reactive chemicals involved, it may require additional help from individuals in research or the lab.

Changes in safety critical instrument test and equipment inspection frequencies

Occasionally, there may be business pressures that would encourage the delay of proof-testing of safety critical alarms and shutdown systems, as well as delay of vessel inspections and delays to safety relief valve testing. There should be some type of variance procedure or review policy defined to handle this occasional need. Such a policy may require the review of all of the inspection and test records on the specific equipment involved as well as an approval of the superintendent of the area.

When hazard studies are performed on new plants and modifications, certain inspection and test frequencies are defined. There should be a review process for changes. If equipment and instrumentation inspections reveal a pattern of failing to meet expectations, the testing or inspection should be increased and alternative equipment must be considered.

There must be an approval mechanism to reduce unnecessary inspection and testing if repeated testing shows ideal performance. Many test frequencies are initially self-imposed to be annual and that could be too frequent. These type of changes should have an approval trail which includes the unit manager, or area supervisor, the critical instrument testing supervisor and a process safety representative.

Approvals, documentation and auditing

Approvals, endorsements and documentation

A responsive modification review and approval system with competent reviewers can gain acceptance quickly. If a modification approval system is unnecessarily cumbersome there can be tension between the sponsor and the reviewer or there can be attempts to circumvent any proposal.

We all must realize that a modification control system, especially for the many small but vital changes, must not be so formal that an answer cannot be expected in a *reasonably short time*. A multiple layer system must be in place to deal with the entire range of proposals from the very simple change to the very complex.

Management of change procedures were reviewed from a number of chemical and petrochemical industries. The typed pages of procedures ranged in length from 4 to 35 pages to fit specific needs. Some were long on definitions and others were long on checklists, but each tried to classify the proposed change to use a flexible approval system.

As mentioned above, minutes should be taken at all chemical process safety meetings. These minutes should be circulated to be reviewed and approved by senior management and senior technical individuals. The minutes of the process safety meetings including the recommendations, the limitations and individuals assigned to handle the follow-up, should be kept for the life of the modification and perhaps the life of the unit involved.

Periodic compliance questionnaires should be sent to the operating unit to check the progress on those recommendations that were made to reduce risk, but were not required to be completed before start-up. It is necessary to verify that recommendations have been completed and this must be acknowledged in the records. Recommendations which were further studied and later deemed impractical or capable of creating additional troubles must not be allowed to remain in limbo. These items should be resubmitted to the process safety committee for re-evaluation.

Auditing the management of change programme

Common sense and the OSHA Chemical Process Safety Management law require an audit of compliance to the 'Management of Change'. Periodic review and documentation of a site's activities in managing aspects of personnel and process safety is often a part of an organization's culture. A good audit can measure the 'actual' versus 'intended' effectiveness of various programmes.

Each organization must devise its own way to conduct an audit. The Dow Chemical Company reported that they have been developing a 'Consolidated Audit' in the Freeport, Texas Plant since 1988. The Consolidated Audit covers safety, loss prevention, occupational health, environment and other topics in a single audit. Prior to 1988, many of these audits were individually achieved on an annual basis.

Dow Chemical was pleased with the efficiency of the combined audit. Since half a dozen audits could be rolled into a single audit, early planning was more practical and more effort could be given to gathering incident records, process flowsheets, and piping and instrument diagrams.

Preparation for the Consolidated Audit began about 3 months before the actual audit, as engineers and plant superintendents reviewed policies and standards and reviewed their plant's status in the areas which were scheduled for audit. A detailed description of their approach was presented at a recent Loss Prevention Symposium (Murphy, 1992).

Other readers may wish to consider details of 'Audits and Corrective Actions' or Chapter 13 of *Plant Guidelines for Technical Management of*

Chemical Process Safety (Center for Chemical Process Safety, 1992a). Chapter 13 covers many of the broad topics of 'scope', 'staffing', 'frequency', 'reports', and 'internal and external auditors'. It is not the intent of this book to cover this aspect of auditing.

Some generic management of change audit questions

These are some commonsense audit questions for the management of change. Other questions can be found in Appendix 13A of *Plant Guidelines for Technical Management of Chemical Process Safety* (Center for Chemical Process Safety, 1992a).

1. Is there a formalized documented policy in place for the review and authorization of changes in the hardware and the operating procedures in units that produce, use, handle, or store hazardous materials?
2. Do all of the affected individuals, including the engineers, supervisors, chemical process operators, maintenance mechanics, purchasing employees, etc., understand that there is a management of change policy?
3. Is there a 'modifications person' available who can provide expertise, has the time to review changes, and can promptly answer process safety questions?
4. Is the system to track and verify process safety recommendations working well and is it up-to-date?
5. Is there a procedure to cope with and authorize 'minor' temporary changes such as operating without some critical alarms and a system to ensure that these 'minor' temporary changes are restored?
6. How is the management of change policy perceived by operators, supervisors, mechanics, engineers, etc.?
7. Have any recent incidents appeared to have been created by a change within the plant, either authorized or unauthorized?

Closing thoughts on a management of change policy

The material in this long chapter may not cover the needs of every chemical plant and every petrochemical plant. But the eight or nine management of change programmes provided to me by major corporations for review did not seem to exactly fit the needs or culture of my organization. Trevor Kletz said many times that improper plant modifications have been a major cause of chemical plant accidents. I have been working in a loss prevention engineering function for the past 18 years and my experiences have been similar. It just seems appropriate to repeat the first three paragraphs of this chapter, as a fitting close.

There must be a formal method to deal with change in the chemical industry. The safety that was designed into the original process often occurs after a multidisciplinary design team agonized over the optimum arrangement of process and layout. This process safety must not be jeopardized by poor-quality modification schemes.

No recipe or procedure can be devised to be universally acceptable. The exact approach used to scrutinize a proposed change must be site specific and developed for that location. There must be a *sustained* management commitment to the management of change programme, since this may require a change in culture within many organizations.

Each chemical plant and refinery must adopt or develop a procedure tailored to fit the specific hazards, the available technical resources, the culture of the organization and required governmental regulations. It must be practical and workable *without undue delays*.

References

American Petroleum Institute (1990) *Management of Process Hazards*, API 750, API, Washington D.C.

Boyen, V.E. (1992) A compliance strategy for process safety regulations. Proceedings of the 26th loss prevention symposium, American Institute of Chemical Engineers, New York

Canadian Chemical Producer's Association (1985) *Essential Components of Safety Assessment Systems*, CCPA, Ottawa, Ontario

Center for Chemical Process Safety (1989) *Guidelines for Technical Management of Chemical Process Safety*, Center for Chemical Process Safety of the American Institute of Chemical Engineers, New York, p. 7

Center for Chemical Process Safety (1992a) *Plant Guidelines for Technical Management of Chemical Process Safety*, Center for Chemical Process Safety of the American Institute of Chemical Engineers, New York

Center for Chemical Process Safety (1992b) *Guidelines for Hazard Evaluation Procedures, Second Edition, with Worked Examples*, prepared by Battelle – Columbus Division, Center for Chemical Process Safety of the American Institute of Chemical Engineers, New York

Chlorine Institute (1990) *Recommendations to Chlor-Alkali Manufacturing Facilities for the Prevention of Chlorine Releases*, Pamphlet 86, Edition 1, CI, Washington D.C.

Heron, P.M. (1976) Accidents caused by plant modifications. *Loss Prevention,* **10,** 72–78

Johnson, R.W. and Leverenz, F.L. Jr (1992) HAZOPS today. Proceedings of the 26th loss prevention symposium, American Institute of Chemical Engineers, New York

Kletz, T.A. (1976) A three-pronged approach to plant modifications. *Loss Prevention,* **10,** 91–98

Kletz, T.A. (1990) *Critical Aspects of Safety and Loss Prevention*, Butterworths, London, pp. 155–157

Lees, F.P. (1980) *Loss Prevention in the Process Industries*, Vol. 1, Butterworths, London, Chapter 8

Murphy, J.F. (1992) Dow Chemical Company's Consolidated Audit, Texas Operations. Proceedings of the 26th loss prevention symposium, American Institute of Chemical Engineers, New York

Organization Resources Counselors, Inc. (1988) *Recommendations for Process Hazards Management of Substances with Catastrophic Potential*, ORC, Washington D.C.

Sanders, R.E. and Wood, J.H. (1991) Don't leave plant safety to chance. *Chemical Engineering* (February), 110–118

US Department of Labor (1992) *Process Safety Management of Highly Hazardous Chemicals; Explosives and Blasting Agents: Final Rule*, OSHA Part 1910.119, US Department of Labor, Occupational Safety and Health Standards, Federal Register, Washington D.C.

Chapter 10

Sources of helpful information when considering modifications

This final chapter lists a number of important resources that should be considered when developing a Management of Change Programme within a chemical plant. Pertinent references were given at the end of each of the previous chapters. However, this chapter is dedicated to providing a list of books, booklets, technical articles and other resources which are essential to consider when addressing chemical plant modifications. Many of these listed references record large numbers of additional references.

These sources will be listed under the following categories:

1. The best five books in chemical process safety - from a process engineer's viewpoint
2. General chemical process safety books
3. Practical Information on aging of pressure vessels, tanks, piping and safety critical instruments
4. Fire and explosion references
5. Other desirable resources

The best five books in chemical process safety – from a process engineer's viewpoint

For those individuals just starting to study chemical process safety and limited reference books, the following titles are strongly recommended. The books are listed in order of projected usefulness, with editorial comments included in some of the entries.

1. Lees, F.P., *Loss Prevention in the Process Industries: Hazard Identification, Assessment and Control*, Volumes 1 and 2, Butterworths, London and Boston, 1980. Containing over 1300 pages, this super reference library is easy to use because it is well written. The large index makes it easy to find specific subject material.
2. *Plant Guidelines for Technical Management of Chemical Process Safety*, American Institute of Chemical Engineers/Center for Chemical Process Safety, (AIChE/CCPS), New York, 1992. This practical easy-to-use book offers specific approaches to various process safety programmes. The book is designed for all chemical process supervision from the foreman to plant management. An entire chapter is devoted to the

'management of change.' This chapter provides detailed policies, checklists, guidelines and assessment sheets.

3. *Management of Process Hazards* – API Recommended Practice 750, American Petroleum Institute, Washington, 1990. It appears that OSHA's November 1991, 'Process Safety Management Law – 1910.119' is based upon this well-worded pamphlet.

4. *Guidelines for Hazard Evaluation Procedures (Second Edition, with Worked Examples)* (AIChE/CCPS), Battelle - Columbus Division, New York, 1992. This document offers excellent overview coverage of evaluation procedures with chapters with titles like: 'Preparing for Hazards Evaluation Studies', 'Hazards Identification Methods and Results', 'Selecting Hazard Evaluation Techniques', etc. Appendix B is excellent. It is titled 'Supplemental Questions for Hazards Evaluation' and contains some of the important questions that are not thought of when using a HazOP. There are 45 pages of probing questions on the process, detailed questions on specific classes of equipment, questions on operation and maintenance and a wide variety of other hazard-searching review questions.

5. Kletz, T.A., *What Went Wrong? Case Histories of Process Plant Disasters*, 2nd edn, Gulf Publishing, Houston, 1988. This is an easy-to-read, popular book. In a few chemical plants, it is required reading for new operations supervisors.

General chemical process safety books

These reference books are listed in alphabetical order and many of the books are followed with a brief comment.

1. Arendt, J.S., Lorenzo, D.K. and Lusby, A.F., *Evaluating Process in the Chemical Industry – A Manager's Guide to Quantitative Risk Assessment*, Chemical Manufacturers' Association, Washington, 1989. This 43-page booklet introduces quantitative risk assessment as an important new tool to complement other historically successful methods for safety assurance, loss prevention and environmental control.

2. Crowl, D.A. and Joseph, F.L., *Chemical Process Safety: Fundamentals with Applications*, Prentice Hall, Englewood Cliffs, New Jersey, 1990. This is the first US college textbook on loss prevention. The textbook has been selling well and is the basis for an AIChE Continuing Education Course.

3. *Essential Components of Safety Assessment Systems*, The Canadian Chemical Producers' Association, Ottawa, Canada, 1985. This 22-page self-assessment form is designed for individual Canadian chemical plants to ensure that the Canadian chemical industry is committed to taking every practical step towards ensuring that its products do not present an unacceptable level of risk to its employees, customers, the public or the environment.

4. *Guidelines for Safe Storage and Handling of High Toxic Hazard*

Materials, Arthur D. Little, Inc. & Richard LeVine for AIChE/CCPS, 1987. This book focuses on good design, fabrication, erection, inspection, monitoring, maintenance, operation, and management practices for high toxic hazard materials. One excellent feature is 8 pages listing helpful tables containing numerous standards, pamphlets and associations which can be contacted for their loss prevention resources. Some of the headings include: 'American Society of Mechanical Engineer (ASME) Codes', 'American Petroleum Institute (API) Publications', 'Selected American National Standards Institute (ANSI) Standards', 'Selected National Fire Protection Association (NFPA) Standards' and other resources.

5. *Guidelines for Technical Management of Chemical Process Safety*, American Institute of Chemical Engineers/Center for Chemical Process Safety, (AIChE/CCPS), New York, 1989. This book describes elements that must be addressed in the development of a technical management system. The *Guidelines* are useful to technical and non-technical managers who have responsibility for managing chemical process technology. This excellent book devotes an entire chapter to the management of change.

6. Kletz, T.A., *Plant Design for Safety – A User-Friendly Approach*, Hemisphere, New York, 1990. This easy-to-read short book is based on the assumption that additional safety can be achieved with safer substitute materials, lower inventories, lower temperatures and lower pressures.

7. Kletz, T.A., *Critical Aspects of Safety and Loss Prevention*, Butterworths, London and Boston, 1990. This book is a collection of nearly 400 thoughts and observations on safety and loss prevention. The topics are generally short and arranged in alphabetical order serving as a sort of 'dictionary of loss prevention'.

8. Kletz, T.A., *Learning From Accidents in Industry*, Butterworths, London and Boston. 1988. The aim of this book is to show, by analysing accidents that have occurred, how we can learn more and thus be better able to prevent similar accidents from occurring. It is another of Trevor Kletz's easy-to-read informative books.

9. Kletz, T.A., *Myths of the Chemical Industry or 44 Things a Chemical Engineer Ought NOT to Know*, IChE, Rugby, UK, 1984. This work makes important points in clever ways.

10. *Loss Prevention*, Volumes 1 to 14, AIChE, New York, 1967 to 1981. These books contain proceedings of the annual Loss Prevention Symposium.

11. *Proceedings of the 24th Annual Loss Prevention Symposium*, AIChE, San Diego, CA, 1990.

12. *Proceedings of the 25th Annual Loss Prevention Symposium*, AIChE, Pittsburth, PA, 1991.

13. *Proceedings of the 26th Annual Loss Prevention Symposium*, AIChE, New Orleans, LA, 1992.

14. *Process Safety Management (Control of Acute Hazards)*, Chemical Manufacturers' Association (CMA), Washington, 1985. This booklet presents a number of ideas on hazard identification, hazard assessment and hazard control in just 47 pages.

Practical information on ageing of pressure vessels, tanks, piping and safety critical instruments

1. Barclay, D.A. Protecting process safety interlocks. *Chemical Engineering Progress*, February 1988, pp. 20–24. This article gives an insight into a maintenance and testing programme within a major plant with over 5000 safety interlocks and alarms protecting hazardous processes.
2. *Inspection of Piping, Tubing, Valves, and Fittings* – Recommended Practice 574, 1st edn, American Petroleum Institute, Washington, 1990. Useful, practical information can be found in this 26-page booklet based on the accumulated knowledge and experience from the petroleum industry.
3. *Pressure Vessel Inspection Code – Maintenance Inspection, Rating, Repair, and Alteration*, API-610, API, 6th edn, Washington, 1989. Another valuable guide, which should be in every significant chemical processing plant, is based upon experiences of participating member petroleum companies.
4. Sanders, R.E. and Wood, J.H., Don't leave plant safety to chance. *Chemical Engineering,* **98**, No. 2, 1991, pp. 110–118. This article offers an overview of desirable elements of a programme to test safety relief valves and safety critical instrumentation.
5. Woolfolk, W. and Sanders, R., Dynamic testing and maintenance of safety relief valves. *Chemical Engineering*, 27 October, 1987, pp. 119–124. This article offers a detailed look at 'in-plant' safety relief valve testing. It includes test facility design, procedures and reporting methods.
6. *Tank Inspection, Repair, Alteration, and Reconstruction*, API-653, 1st edn, API, Washington, 1991. Another very useful guide that is a necessity in refineries and chemical plants.

Fire and explosion references

1. Bodurtha, F.T., *Industrial Explosion Prevention and Protection*, McGraw-Hill, New York, 1980.
2. *Dow's Fire and Explosion Index, Hazard Classification Guide*, 5th edn, A Chemical Engineering Progress Technical Manual, AIChE, New York, 1981. This booklet provides excellent, practical classification in 57 pages.
3. Vervalan, C.H., *Fire Protection Manual for Hydrocarbon Processing Plants*, Gulf Publishing, Houston, 1985.
4. Zabetakis, M.G., *Flammability Characteristics of Combustible Gases and Vapors*, US Bureau of Mines Bulletin 627, US Department of the Interior, Washington, D.C., 1965.

Other helpful resources

1. The American Institute of Chemical Engineers (Center for Chemical Process Safety) has announced a computer-based training module

entitled 'Management of Change'. This module uses interactive incident scenarios to walk participants through basic plant modifications. Users are exposed to scenarios and narratives that demonstrate six basic types of change. The exercises vary for each type of job, such as plant operations or maintenance. This computer-based module was expected to be released in December 1992: it has not been reviewed by this author.

2. The American Institute of Chemical Engineers (AIChE) sponsors over a dozen 'Continuing Education Courses' which focus on process loss prevention. Each of these courses is usually offered two or three times a year. A partial list of these courses include:

   Chemical Plant Accidents – a Workshop on Causes and Prevention
   Fundamental of Chemical Process Safety
   Fundamentals of Fire and Explosion Hazards Evaluation
   Loss Prevention Management
   Use of Hazard and Operability Studies in Process Analysis

   The AIChE also sponsors an annual Loss Prevention Symposium and the Center for Chemical Process Safety. Further information can be obtained from:

   American Institute of Chemical Engineers
   345 East 47th Street
   New York, NY 10017 USA
   Telephone (212) 705-7338

3. The American Petroleum Institute (API) maintains cooperation between US government and industry on all petroleum-related issues of US concern. The API provides process hazards and process safety seminars, technical standards of design and fire protection, and equipment inspection guides which are applicable in most plants. Specification information can be obtained from:

   American Petroleum Institute
   1220 L St, N.W.
   Washington, D.C. 20005, USA
   Telephone (202) 682-8000

4. The American Society of Mechanical Engineers (ASME) provides the boiler and pressure vessel codes which are widely accepted as the basis for construction of boilers and pressure vessels. Further information can be obtained from:

   American Society of Mechanical Engineers
   United Engineering Building
   345 East 47th Street
   New York, NY 10017 USA
   Telephone (212) 705-7800

5. The Chemical Manufacturers' Association (CMA) is a trade association of chemical manufacturers, who produce more than 90% of the basic industrial chemicals in the United States. The CMA offers a national chemical referral and information centre as well as a number of

of management guides for risk assessment and process safety. The CMA can be contacted by telephoning or writing to:

Chemical Manufacturers' Association
2501 M St, N.W.
Washington, D.C. 20005, USA
Telephone (202) 887-1100

6. The Compressed Gas Association (CGA) is composed of members from firms that produce and distribute compressed, liquefied and cryogenic gases. The association develops and publishes recommendations to improve safety concerning the transportation, storing and handling of gases. The CGA provides handbooks, technical publications, videotapes, and safety posters relating to safety. The CGA can be contacted by writing to or calling:

Compressed Gas Association
1235 Jefferson Davis Highway
Arlington, VA 22202, US
Telephone (703) 979-0900

7. The Institution of Chemical Engineers (IChemE) offers the 'Loss Prevention Bulletin – Articles and Case Histories from Process Industries Throughout the World' (Bernard M. Hancock, editor). This Bulletin is published every second month and is available on a subscription basis. Further information may be obtained from:

Bernard M. Hancock
Manager – Safety and Loss Prevention
The Institution of Chemical Engineers
165–171 Railway Terrace
Rugby, CV21 3HQ
UK
Telephone (0788) 578214 or FAX (0788) 560833

The Institution of Chemical Engineers also offers a full range of training materials in the form of Hazard Workshop Training Modules. These modules comprise case studies including documented accidents, 35 mm colour slides and notes for the discussion leader. Several of the modules are supplied with a video in the package. Currently, there are 17 of these interactive training modules available for purchase. Some of the more popular modules are listed:

Hazards of Over- and Under-pressuring Vessels
Hazards of Plant Modifications
Fires and Explosions
Human Error
Inherent Safety
Safer Piping

For worldwide distribution, contact Bernard Hancock.

For North and South America, contact:

Dr J.C. Forman
Forman Associates Consulting and Technical Services
77 Stanton Road
Darien, CT 06820, USA
Telephone (203) 655-4180 or FAX (203) 655-0789

8. The National Fire Protection Association (NFPA) has developed, published and disseminated a large number of practical standards to minimize the possibility and effects of fires and explosions. The NFPA also produces reference books, test books, posters and slide/audiotape programmes on fire concerns. A list of a few of the more popular consensus standards related to fire and explosion would include:

30 – Flammable and Combustible Liquids Code
49 – Hazardous Chemical Data
58 – Liquefied Petroleum Gases, Storage and Handling
68 – Explosion Venting
70 – National Electric Code
325M – Fire Hazard Properties of Flammable Liquids, Gases and Volatile Solids

The NFPA can be contacted at:
National Fire Protection Association
Batterymarch Park
Quincy, MA 02269, USA
Telephone: (617) 770-3000

9. *1991 CCPS/AIChE Director of Chemical Process Safety Services*, AIChE, New York, NY 1991. This directory contains a compilation of more than 350 sources of chemical process safety services offered within the United States and Canada. The book contains the names and addresses of companies offering consulting services, emergency services, testing assistance and training schools.

Index

Abrasive materials in pipes, 113
Accidents, AIChE course on, *xii*
Acid,
    cleaning by, 35
    damage to tank, 8–10
Acoustic emission testing, 111
Acrylic, acid attack on, 69
Actual test, 122
AIChE (American Institute of Chemical
    Engineers), *xii*, 81, 115, 164
    (*see also* CCPS)
    continuing education, 159, 162
    guidelines, 148
    information from, 149
    loss prevention symposium, 123, 135
    meeting, 115
Air, back-up, disconnection, 64
Air drier explosion, 19–23
Air-line,
    ammonia in, 78
    water in, 51
Air-rolling, 33
Algae, elimination of, 75–6
Alkali, cleaning by, 35
American Institute of Chemical Engineers,
    see AIChE
American Insurance Association, 58
American Petroleum Institute *see* API
American Society of Mechanical Engineers
    see ASME
Ammonia,
    cleaning problems, 37
    contamination by, 78
    explosion, 34
ANSI standards, 160
ANSI codes, 106
Antifreeze, 75
API (American Petroleum Institute), 37
    government–industrial liasion, 162
    information from, 149
    literature, 160
    publications on inspect, 108
API codes for atmospheric and low-pressure
    storage tanks, 106

API750, 138
Approvals, 154
Asbestos gasket, 67–9, 96
Asbestosis, 68
ASME (American Society of Mechanical
    Engineers) codes, 106, 111, 160, 162
Audit, 154–6
    consolidated, 155
    modification procedure, 133
Authorization, 136
    for changes in job scope, 56
    need for, 105
    nonroutine work, 37

BCF, 99, 100
Benzene fire, 96
Bhopal incident, 79, 80–1
BLEVE (boiling liquid expanding vapour
    explosion), 60, 61
Blinds,
    listing, 134
    problems due to, 51–5
BMA (British Medical Association), 6
Boiling in pipes, 113
BP chemicals international, 135
Brass as pipeline hazard, 89
Breathing, compressed air used for, 50
Brine pump, 94
Brine sludge, dissolving system for, 30–2
British Medical Association, 6
Bromine, release of, 75
Building modification, damage from, 97
Butane, 60

Cadmium for priming steel, 108
Calandra, A. 103–4
Canadian Chemical Producers Association
    see CCPA
Cancer, asbestos and, 68
Capillary action, 107
Carbon build-up, 85

Carcinogenic materials, 68, 106–7
Case histories, *xii*, 3, 159
Cash resources, 144
Cast iron,
  brittle fracture of, 95
  as pipeline hazard, 89
  pump bowl, 94
Catalyst pipes, 113
Cause-and-effect approach, 124
Caustic soda,
  and aluminium, 97
  cleaning by, 35
  corrosion by, 33, 115
  dilution with water, 32–3
  icicles, 65, 66
CCPA (Canadian Chemical Producers
    Association), on safety assessment, 136
CCPS (Center for Chemical Process Safety),
    131, 138, 164
  computer-based training, 161
  guidelines, 141, 147
Celluloid plastics, development of, 2
Center for Chemical Process Safety, see
    CCPS
Changes, *see* Modifications, 140
Chemical engineering, 114, 116, 124
Chemical engineering, progress in, 81
Chemical Manufacturer's Association, *see*
    CMA
Chemical reaction, unforeseen, 30–2
Chlorine
  escape, 64
  Institute, 35, 139
  increasing supply of, 79–80
  release of, 75
Clamp fracture, 55–6
Clampdown table, 116
Cleaning,
  chemical, 35–7
  safety problems, 37
CMA (Chemical Manufacturers'
    Association), 105, 115, 162
  risk manual, 5
Coal-tar dyes, development of, 2
Coatings, inspection of, 109
Coker drum, failure of, 13–16
Combustible materials, *xii*
  equipment integrity for, 106–7
Commitment by management, 142
Communication,
  need for, 105
  problem, 41
Compliance questionnaire, 155
Compressed air, breathing of, 50
Compressed Gas Association (CGA), 163
Compressor, centrifugal, 90
Computerized maintenance, 130
Computerized test results, 130
Condensation in pipes, 113
Consolidated audit, 155
Containment integrity, 106

Contingency planning, need for, 105
Continuity vs. safety, 105–6
Contractor safety, 41
Cooling tower, alarm/trip for, 129
Cooling water system, explosion caused by,
    75
Corrosion,
  by caustic soda, 33
  failure, 60
  ferric chloride, 90, 92
  hazards, 93
  inspection for, 108
  insulation, 107–8, 113
  patterns, 110, 111
  piping, 37–9
  stress cracking, chloride, 108
  threat, 90
Custom fabrication purchasing agent, 111
Cyclohexane escape, 79

Dangers, exaggeration of, 99
Death expectancy, accidental, 6–7
Demand rate, 129
DEMCO valve, 49
Deviations, 152
Documentation, 140, 154
  of modifications, 133, 137
  need for, 105
Dow Chemical Company, 155
  fire and explosion index, 150
  inspection programme, 111–13
Drain, 150
Drugs, development of, 2
Dupont Company, 122–3
  alarm/trip for, 129
Dye penetrate examination, 111

Earthing, inspection of, 109
Eddy current methods, 112
Electrical arcing, ignition by, 34
Emergency planning, 137
Environmental release, 129
Environmentally objectionable materials,
    equipment integrity for, 106–7
Epoxy coatings,
  amine-cured tar, 108
  catalysed, 108
  phenolic, 108
Erosion hazards, 93
Ethane, 59
Ethylene,
  glycol, 75
  polyethylene formation, 48
  tank, 23–5
Evaluation studies, 159
Exothermic reaction vessel, 127
Expansion bottle, 35, 36
Expansion joint, 90
Experience dilution, 136

Explosion,
    information sources, 161, 163
    relief, 87
    vent, information on, 131
External inspection, 108, 109
Exxon Chemical Americas, 105

FAFR (fatal accident frequency rates), 5–7
Fatal accidents, 4–7
Fault tree analysis, 137, 149, 151
Feedstock supplies, 60
Fermentation, 1
Fibre, 3
Fire, 163
    damage, 16–17
    information sources, 161
Fireproofing, inspection of, 109
Flame failure, 127
Flammable liquid transfer, 100
Flammable materials, *xii*, 106–7
Flanges, inspection of, 109
Flash point explosion, 43, 45
Flixborough incident, 57–8, 79
Food processing, safety record, 34
Foundation, inspection of, 108, 109
Freezing hazards, 24, 44, 52, 85
    during steam cleaning, 37
Fuel, 3
    shutdown system, 127
Function test, 122
Furnaces, abuse of, 86
Fuse plug, accidental melting, 80

Gasket, 65–9
    acrylic, 69
    asbestos, 67–9, 96
    elastomers for, 115
    graphite, 68
    piping, 68–9
    Teflon, 68
    wrong, 65, 67
Glass, 1, 2
Class fibre, 3, 108
Glycerin, nitroglycerin from, 63   ·
Government,
    inspection by, 109
    legislation, 138
Grounding, inspection of, 109
Gulf Coast chemical plants, 69, 83, 107–8

Halon, 99, 100
Hancock, B.M., 163
Hazard rate, 129
Hazards, *xii*, 3
    (evaluation, 159)
    identification of, 148, 159
    prediction of, 150
    studies, 149
    training kit, 11, 25, 163
    reviews, 137

HazOp, 150, 151, 159
Head-count, 134
Heads, inspection of, 109
Health and safety executive, 6
Heat of solution, 37
Heron, P., 135
History of chemical industry, 1–2
Hose,
    contamination in, 78
    misuse of, 78–81
    warming by, 79
Hot work permits, 41
Human error, 163
Hurricane threat, 69
Hydrazine contamination, 78
Hydrocarbon–chemical industries, damage
        losses, 59
Hydrogen chloride, corrosion by, 115
Hydrogen line failure, 59
Hydrogen–oil fire, 67
Hydrogen sulphide, release during cleaning,
        36
Hydrostatic test, 41–42

ICI, 99
    modification procedure, 136
    Mond Division, 150
Identifying tags, 72, 153
'Imagine-if' modifications, 99, 149
Information sources, 158–64
Inspection,
    external, 108, 109
    frequency, 109
    internal, 110
    low-pressure tank, 111
    planning of, 106
    routine changes, 154
    unnecessary, 154
Institution of chemical engineers (IChemE),
        11, 58, 163
    awareness training, 93
    fires and explosions, 86
    on hazards, 29, 141
    information from, 149
    vaporization expansion, 42
Instrument,
    critical, 126
    out-of-service, 153
    priorities, 152
    setting changes, 154
    specification, 134
Instrument air, alarm/trip for, 129
    blockage clearance by, 80
Instrumentation,
    modification of, 64
    safety improvements, 65
Insulation,
    corrosion under, 107
    inspection of, 108, 109
Insurance guidelines, 106
Internal inspection, 110

Iron-in-chlorine fire, 45–6
Isobutane explosion, 37, 47

Joseck, J. L., 99

Kettle alarm/trip for, 129
Kletz, T. A., xi, xii, 41, 134, 155
    dangers, exaggeration of, 99
    hazard workshop, 11
    on HazOp, 150
    on modifications, hazards of, 136
    safety checklist, 149
Knowledge dilution, 134–5

Law, US, on hazardous chemicals, 83
Leblanc, N., 2
Lees, F., 148
Legionnaire's disease, 75
Lightweight materials, 3
Lining, inspection of, 108
Liquid metal embrittlement, 69
Loss prevention association, Louisiana, 115
Loss prevention bulletin, 75
Loss prevention symposium, 81, 123, 135,
    155, 160
Loss-of-time injury, 4
Low-pressure tank, inspection of, 111
Lpg pipeline, 61
Lubricant, wrong, 63–4
Lung pressure, 30

Magnetic-particle examination, 111
Maintenance,
    costs, 28
    procedures, dangers of, 28–41
    routine, 28
Mansfield, D., 75
Marsh & McLennan Inc., 59
McGraw-Hill publications, 114, 124
McMillan, G., 123
Mechanical integrity, 41, 131
Metal thickness variation, 110
Mist, combustible, 43
Modifications,
    approval for, 154
    control of, 133–57
    decisions on, 142, 143
    hasty, 63–81, 134
    identifying consequences, 136
    information sources, 158–64
    person, 141, 146
    prerequisites for, 142
    rushed, 63–81, 134, 137
    side effects of, 8–26
    significant vs. trivial, 142–4
    temporary, 140
Monsanto Chemical Company, 123
Munitions, 3

NACE publication 6H189, 108
NACE (National Association of Corrosion
        Engineers),
    anti-corrosion coats, 108
    chlorides, 107
    corrosion under insulation, 107, 108
National Association of Corrosion
        Engineers, see NACE
National Fire Protection Association, see
        NFPA
National Safety Council, 4, 6, 34
Neoprene, 3
NFPA (National Fire Protection
        Association), 106, 149, 160, 164
Nitrates in cooling water, 79
Nitric acid gas, 63
Nitrogen, oxides of, 19
Nitroglycerin from glycerin, 63
Nitroglycerin explosives, development of, 2
Nitrous acid gas, 63
Nozzles, inspection of, 109
Nuclear power, growth slowed, 80
Nypro (UK) Ltd, 57, 79

O-rings, elastomers for, 115
Obsolete facilities, 137
Occupational Safety and Health
        Administration, see OSHA
Oil explosion, 43
    flash point of, 43
    hydrogen fire, 67
    vapour hazard, 42
Oil mist fire, 88
One-minute modifiers, 63–81, 134, 137
Operability studies, 149
Operating parameters, 139
Operator aid, 122
ORC (Organization Resources Counselors,
        Inc.), 138
Organizations, list of, 161–9
OSHA (Occupational Safety and Health
        Administration), xii, 38
    compliance guidelines, 139
    final rule, 141
    law, 155
    Phillips 66 incident, 47
    piping arrangement, 49, 50
    safety law, 139
    safety management, 41
Overfilling, tank damage by, 8–13
Overpressuring, 163

Packaged units, hazards of, 97
Painters, problems caused by, 70–3
Paperwork, overcomplicated, 134, 147
Petrol fill-up example, 101
Phenol explosion, 44
Phillips 66 incident, 47–9, 60
Pipefitting, problems caused by, 72–4

Piping, changes to, 87–93
  construction materials, 58
  corrosion of, 37–9, 58
  dead legs, 58
  design failure, 60
  expansion problems, 89
  hazards, 58–61, 89
  inspection of, 108, 113
  overpressure, 58
  routing, 58
  safer, 163
  specifications ignored, 87–93
  spool incident, 46
  support, 58, 89, 90
  thermal expansion, 58
  threaded, 58, 89
  vibration, 58
Plant inspection and maintenance forum, 105
Plastic pump and flammable liquids, 95
Pleural plaques, 68
Pollution control, fire/explosion risk, 16,
  19–22
Polyethylene, 3
  from ethylene, 48
Pop SRV testing, 116
Power, alarm/trip for, 129
PPG industries, 115, 116, 126, 129
Pressure,
  alarm/trip for, 129
  build-up fluids, 34–5
  safety, control of, 127
  vessel codes (ASME), 162
  vessel, inspection, 161
Pressure gauge, 110
Pressure testing, 111
Pressure vessel, inspection of, 108, 109–11
Pressure Vessel Review Committee, 111
Pressure Vessel Review Section, 112
Problem solving, practical, 102–4
Proof-testing, 65, 124–30
Propane, 59, 60
Protective programmes, 105
Pump, hazards with, 94–5
Purging, safety problems, 37
PVRS (Pressure Vessel Review Section), 112

Quality assurance, 106

Radiography, 111, 112
Reactor collapse, 59
Rectifier, alarm/trip for, 129
Refrigerated tank, 23–5
Refrigeration compressor, alarm/trip for,
  129
Regulations, revised, 138
Relief valve, 35, 72
Repair procedure, 134
Replacement in kind, 139, 140, 142–3
Risk assessment, 100–1, 159
Robertson, K., 105
Roof, easy separation for safety, 86, 87

Rupture disc, 35, 110, 111
  information on, 131
  wrong installation, 74

Safeguard failures, record of, 70
Safer piping module, 58
Safety, 3–5
  alarms, 122
  assessment, 136–7
  checklists, 145–7, 149
  devices, 124, 127–9, 152–3, 161
  guidelines, 146
  inherent, 163
  interlocks, 122–3
  literature, 158–64
  management, 100, 138
  meetings, minutes of, 155
  newsletter (ICI), 99
  record, food processing, 34
  relief valve, *see* SRV
  review team, 147
  self-assessment, 159
  systems, assurance of, 114
  training, 93
  vs. continuity, 105–6
Sandblasting, problems caused by, 70
Sealing, unnecessary/damaging, 69–71
Service line, 150
Shear wave ultrasonics, 112
Shell, inspection of, 109
Sight glass, obscured, 72
Simmer SRV testing, 116
Simulated test, 122
Skirts, inspection of, 109
Slurry piping, 113
Soap, 2
Soda ash, 1–2
Sodium chloride, corrosion by, 115
Solution, heat of, 37
Specifications, need for, 83, 134
Spring support, 90
Sprinkler, 72, 127
SRV (safety relief valve), 72, 73
  function of, 114–15
  identification, 121
  information on, 131
  inspection of, 109
  records, 121
  repair, 118–20
  testing, 115, 116, 117–21
Static electricity, *xii*
Steaming-out, safety problems, 37
Still, alarm/trip for, 129
Storage tank,
  damage to, 8–13
  inspection of, 108, 109–11
  low pressure, 86
Stress cracking, 108
Stud-bolt failure, 69
Styrene plant incident, 96

Sulphuric acid leak, 37
Supports, inspection of, 109
Synthetic rubber, 3

Tagging checklist, 153
Tank,
    accessories, inspection of, 109
    low-pressure, inspection of, 109
    problems with, 8–26
Temperature,
    alarm/trip, 64–5, 129
    safety, control of, 127
Test procedures, communication of, 121
Testing, limited-volume, 116
Three-Mile Island incident, 79, 80
TOP event, 151
Toxic materials, *xii*, 106–7, 159–60
Training,
    computer-based, 161
    overview, 141
    program, 136
Turbulence and metal loss in pipes, 113

Ultrasonic thickness measurement, 112
Underpressuring, 163
Union Carbide India Ltd, 80
US Code of Federal Regulations, 139

US Department of Labor, *xii*, 47
US Regulators, maintenance, 41

Vacuum, damage by, 8–16, 28–30
Valve,
    emergency isolation, 127, 128
    inspection of, 109
    safety relief, *see* SRV
Vaporization hazards, 16–17, 42, 137
Vapour, toxic release of, 64
Velocity and metal loss in pipes, 113
Vent, emergency, information on, 131
Vent modification, hazards of, 75–8
Vessels, repairs, revisions, status changes,
    112
Vinyl chloride, escape of, 46
Vulnerability, identification of, 148

Wall Street Journal, 80
Walls, lightweight for safety, 96
Washing, safety problems in, 37
Water, intruding, 99, 107
Wax blockage, 76–7
Welding, ignition by, 43–5
'What if' studies, 99, 149
Wool, introduction of, 102
Woolfolk, W., 116

Zinc for priming steel, 107, 108